# SKILLS Coach
## America's Best for Student Success

# Write Math!

### How to Construct Responses to Open-Ended Math Questions

2nd Edition

LEVEL G

Andrew Kaplan

Write Math!
How to Construct Responses to Open-Ended Math Questions, Level G
102NA
ISBN-10: 1-58620-913-2
ISBN-13: 978-1-58620-913-1

EVP, Publisher: Linda Sanford
VP of Production: Dina Goren
VP, Creative Director: Rosanne Guararra
Art Director: Farzana Razak

Senior Development Editor: Elizabeth Jaffe
Contributing Author: Keith Grober
Design: Electric Pictures
Layout artist: Rocio Paez
Illustrator: Jacob Nicholas
Cover Design: Farzana Razak
Cover Photo: Myron/The Image Bank/Getty Images

**Triumph Learning**® 136 Madison Avenue, 7th Floor, New York, NY 10016
© 2006 Triumph Learning, LLC
A Haights Cross Communications, Inc. company

All rights reserved. No part of this publication may be reproduced in
whole or in part, stored in a retrieval system, or transmitted in any form
or by any means, electronic, mechanical, photocopying, recording or
otherwise, without written permission from the publisher.

Printed in the United States of America.

10 9 8 7 6 5 4 3

# Table of Contents

Letter to the Student .................................................. 5

**Chapter One:** What Is an Open-Ended Math Question? ..................... 6

**Chapter Two:** What Is a Rubric? ....................................... 12

**Chapter Three:** How to Answer an Open-Ended Math Question .............. 18

**Chapter Four:** How NOT to Get a Zero! ................................. 24

**Chapter Five:** Number and Operations .................................. 30

**Chapter Six:** Algebra ................................................. 48

**Chapter Seven:** Geometry .............................................. 70

**Chapter Eight:** Measurement ........................................... 92

**Chapter Nine:** Data Analysis and Probability ......................... 110

**Chapter Ten:** Tests .................................................. 128

**Chapter Eleven:** Home-School Connection .............................. 142

Glossary .............................................................. 149

Math Reference Charts ................................................. 168

# Letter to the Student

**Dear Student,**

Welcome to the smart way to write answers to open-ended math questions. You will learn what open-ended math questions are, how to solve them, and how to score them. You will do this by reviewing modeled problems, practicing with guided questions, and answering independent problems. You will work together with your teacher, with your classmates, and with your caregiver at home.

Let's learn to **write math** the smart way!

**Have fun!**

Mathematics Open-Ended Questions, Level G

# 1. What Is an Open-Ended Math Question?

Just like other math problems, an open-ended math problem has a correct answer. However, in an open-ended math problem, you can use more than one method to find this answer. Any method you use is fine, as long as:

- it produces the correct answer.
- you show how it got you to the answer.
- you explain the strategy/strategies you used.

NOTICE: Photocopying any part of this book is prohibited by law.

# 1. What Is an Open-Ended Math Question?

**5-step plan:**

**1. Read and Think** • **2. Select a Strategy** • **3. Solve** • **4. Write/Explain** • **5. Reflect**

This plan will help you answer open-ended questions.

1. Read and Think
2. Select a Strategy
3. Solve
4. Write/Explain
5. Reflect

**1. Read and Think**

Read the **problem** carefully.

What **question** are you asked?

- In your own words, tell what this problem is about.

What are the **keywords**?

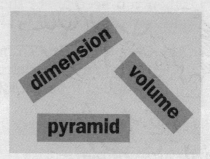

What **facts** are you given?

- Decide what facts are needed, and which ones are extra.

NOTICE: Photocopying any part of this book is prohibited by law.

# 1. What Is an Open-Ended Math Question?

## 2. Select a Strategy

- How am I going to solve this problem?
- What strategy should I use?

There are lots of strategies to help you solve open-ended math questions. Some are listed below and others you may come up with yourself. In parentheses, you will find places where these strategies are used in this book.

### 1. Draw a Picture or Graph . . .

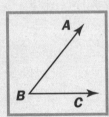

when you need to see the information given in a problem.

(see Chapter 6, Guided Problem #2, p. 60; Chapter 7, Modeled Problem, p. 71, Guided Problem #2, p. 83)

### 2. Make a Model or Act It Out . . .

when you need to watch how the solution is found.

(see Chapter 8, Guided Problem #1, pp. 97-98; Chapter 9, Guided Problem #1, p. 114)

### 3. Make an Organized List or Table . . .

1. \_\_\_\_\_
2. \_\_\_\_\_
3. \_\_\_\_\_

when you need to organize information in a problem. A list or table can help you relate different sets of data and discover patterns.

(see Chapter 5, Guided Problem #1, p. 35; Chapter 6, Guided Problem #3, p. 64; Chapter 7, Guided Problem #2, p. 82; Chapter 9, Guided Problem #3, p. 124)

### 4. Look For a Pattern . . .

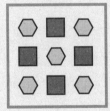

when you need to predict what comes next or find a rule. Making a list or table may help you find a pattern.

(see Chapter 6, Modeled Problem, p. 50, Guided Problem #3, p. 64); Chapter 9, Guided problem #2, p. 119.

### 5. Guess and Test . . .

when it is difficult to find the answer to a problem, or when you want to test an idea about how to solve the problem. Make a guess. Then test it. If your guess is incorrect, use that guess to make a better guess.

(see Chapter 7, Guided Problem #1, p. 77; Chapter 9, Guided Problem #3, p. 124)

NOTICE: Photocopying any part of this book is prohibited by law.

# 1. What Is an Open-Ended Math Question?

### 6. Logical Thinking . . .

when you need to figure out how the information you have fits together.

(see Chapter 5, Guided Problem #2, p. 39; Chapter 8, Guided Problems #2, p. 101, #3, p. 105-106; Chapter 9, Modeled Problem, p. 111)

### 7. Work Backward . . .

when you know the end result or total, and need to find a missing part.

(see Chapter 5, Modeled Problem, p. 31)

### 8. Write a Number Sentence or Algebraic Equation . . .

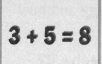

when you need to show your calculations or find an unknown quantity. Often, in order to solve a problem, you need to translate a problem situation into an algebraic equation in which a variable like $n$ represents an unknown number.

(see Chapter 6, Guided Problems #1, p. 154, #2, p. 59; Chapter 9, Guided Problem #1, p. 115)

### 9. Divide and Conquer . . .

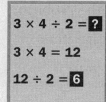

when you must solve more than one equation to find the answer to the main problem. You can also use this strategy to solve a problem that does not involve calculations. Break the problem into steps, and solve one step at a time.

BE SURE to explain how you solve each step and what strategy/strategies you use to solve each step.

(see Chapter 5, Guided Problems #2, p. 39, #3, p. 43; Chapter 7, Guided Problem #3, p. 87; Chapter 8, Modeled Problem, p. 93)

### 10. Make It Simpler . . .

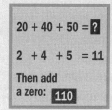

when you must solve a complex problem with large numbers or many items.

Reduce the large numbers to small numbers, or reduce the number of items given. You can also use a pattern from a simpler problem to find an answer to a more complex problem.

(see Chapter 7, Guided Problem #3, p. 87; Chapter 8, Guided Problem #2, p. 101)

NOTICE: Photocopying any part of this book is prohibited by law.

# 1. What Is an Open-Ended Math Question?

### 3. Solve

After you pick your strategy, use it to solve the problem.

Use your mathematical skills and knowledge here.

- Be careful not to make a mistake in your calculations.

- Be careful—you want to get the correct answer. Make sure you are using the facts. If you use a formula, make sure it's the correct formula. Write all of the steps. Check your calculations!

### 4. Write/Explain

Write out an **explanation** of how you solved the problem.

Explain the strategy you chose and why you chose it.

Write about how you solved the problem and why you chose to solve it that way.

Don't leave out any steps.

- If you came up with a strategy of your own that is not on the list on pages 8–9, be sure to explain what the strategy is and why you chose it!

- Your writing must be clear. This is very important when you are taking a test. Remember, the person who reads your work must be able to figure out what you did. You can lose points if your writing is not very clear.

### 5. Reflect

**First, Review Your Work**

Review what you have written using this list.

**Read It and Think**

- ☐ Did I read the problem at least twice? Do I understand it?
- ☐ Did I write down the question being asked?
- ☐ Did I write down the keywords in the problem?
- ☐ Did I write down the facts that are given?
- ☐ Did I write down the strategy that I used?
- ☐ Did I solve the problem?
- ☐ Is my arithmetic correct? Are my calculations correct?
- ☐ Did I explain how I solved the problem?
- ☐ Did I explain why I chose the strategy and how I used it?
- ☐ Did I include all the steps I took to solve it?
- ☐ Is my writing clear?
- ☐ Did I label my work?
- ☐ Does my answer make sense?
- ☐ Did I answer the exact question being asked?

**Always check your work!**

NOTICE: Photocopying any part of this book is prohibited by law.

# 1. What Is an Open-Ended Math Question?

## Then, Improve What You Wrote

How can you improve your writing?

- Try to rewrite your answer to make it clearer, more accurate, and complete.

These five steps you just reviewed might seem like a lot of work. You may not think that some of them are necessary. But after you use them for a while, they will become habits for you. These are the habits of a successful problem solver who gets good scores. You will find that these five steps will work with any open-ended math question. This book will help you practice using them.

Working with this book will help you do better on math tests with open-ended questions. You will even learn how to check your answer using a rubric on *page 13*.

## The Glossary

The Glossary of this book is found on *pages 149–167*. It contains the meanings of mathematical words, including the **keywords** found in the problems throughout this book.

Sometimes a word can mean one thing in everyday life, but something else in mathematics. For example, in reading, the word **supplementary** can mean *additional*. "The **supplementary** material helped me understand the information in the textbook more easily." But in geometry, two angles are **supplementary** if they have a sum of 180°.

If you are not sure of the mathematical meaning of a word in this book, look it up in the Glossary.

### Tips

- If you came up with a strategy of your own that is not listed on *pages 8-9*, be sure to explain what the strategy is and why you chose it!

- Your writing must be clear. This is very important when you are taking the test.

- Remember, the person who reads your work must be able to figure out what you did. You can lose points if your writing is not very clear.

NOTICE: Photocopying any part of this book is prohibited by law.

**Mathematics Open-Ended Questions, Level G**

# 2. What Is a Rubric?

A **rubric** is a grading system used to score **open-ended math questions.** The person who scores the answers on your test uses a rubric. A rubric can also be used as a guide in answering open-ended math questions. It lists the things that should be found in your answer for a high score. It also describes answers that are less than complete that would give you a low score. You can also use a rubric to review and improve your own work.

Throughout this book, you will use the rubric in two ways:

1. to **guide** you in answering the open-ended math question. It works like a checklist, reminding you to write a correct, clear, complete, and thoughtful answer.

2. to **score** yourself as you double-check that your answer is complete.

NOTICE: Photocopying any part of this book is prohibited by law.

## 2. What Is a Rubric?

Here is a typical rubric. It is used to score work from **0 to 4**. **4** is a perfect answer.

**4**
- You showed you knew what the problem asked, including keywords.
- You showed you knew what facts were given, including keywords.
- You chose a good strategy and used it correctly.
- Your arithmetic or operations were done correctly.
- You got a correct and complete answer and labeled it.
- You wrote a good, clear explanation of why you chose a strategy and how you used it.
- You put in all the steps you used to get to your answer.
- You explained your thinking clearly.

**3**
- You showed you knew what the problem asked.
- You showed you knew what facts were given, including keywords.
- You chose a good strategy but may not have used it correctly, OR you may have made an arithmetic error in your work.
- You wrote an explanation of why you chose a strategy and how you used it.
- You might not have used all of the steps to get your answer.
- Your explanation was mostly clear but might not have been entirely complete.

**2**
- You showed you knew what the problem asked.
- You showed you knew what facts were given, including keywords.
- You chose a good strategy but may not have used it correctly, OR you may have made an arithmetic error in your work.
- Your answer may not be correct.
- Your explanation may not be complete.
- Your explanation may not be clearly written.

**1**
- You did not understand what the problem asked, OR you did not know what facts were given.
- You did not select a good strategy or did not apply your strategy correctly.
- You made an arithmetic error in your work.
- Your explanation was not complete or you did not write an explanation.
- Your explanation was not clearly written.

**0**
- You showed no work at all, OR the work you showed had nothing to do with the problem.

### Tips

- You will never get a score of **0** if you start to solve the problem.

- You should always write down what you were asked and what facts were given. This shows that you understood some of the problem, and attempted to solve it.

- Using a rubric may seem like a lot of work. You may not see why it is necessary. But once you use it to answer a few problems, you will see how it can help you improve your answers and your scores. You should practice using a rubric at school and at home.

NOTICE: Photocopying any part of this book is prohibited by law.

# 2. What Is a Rubric?

Let's do an open-ended math question. See how the **5-step plan** fits the rubric to help **YOU** get a 4 on your next answer!

Remember the **5-step plan**:

1. Read and Think • 2. Select a Strategy
3. Solve • 4. Write/Explain • 5. Reflect

### 1. Read and Think

Read the **problem** carefully.

### Modeled Problem

Roberta has had a monthly membership in The Northside Fitness Center for 15 months. For a **monthly** membership, the fitness center charges a one-time registration fee of $150 and an additional monthly fee. So far, Roberta has spent a total of $1,350. What **amount** is the monthly fee?

**Keywords: monthly, amount**

What **question** are we asked?

- The question tells you what it is you want to find out.
- The question is "What is the monthly fee?"

What are the **keywords**?

- **monthly** happening once a month
- **amount** value, number, or

What **facts** are we given?

- Every problem has facts, data, or information that help you answer the question.

Check the glossary on page 149.

In this problem, the **facts** are:

- Roberta has belonged to the fitness center for 15 months.
- There is a one-time registration fee of $150 and an additional monthly fee.
- Roberta has spent a total of $1,350.

### 2. Select a Strategy

In order to solve a problem, you need to use a **strategy**.

## 2. What Is a Rubric?

There are many strategies you can use. In *Chapter 1*, *pages 8–9* show some strategies you might use. You may also choose to use one of your own.

Look at how two students chose different strategies to solve this problem.

### First Solution

**2. Select a Strategy**

First, we'll take a look at how Janine used algebra to solve the problem. This involves using a strategy, **Write an Algebraic Equation.**

**3. Solve**

First solution—**Write an Algebraic Equation.**

$n$ = monthly fee
$15n$ = monthly fees for 15 months
registration fee + monthly fees for 15 months = total
Janine pays in 15 months
$150 + 15n = 1,350$
$150 - 150 + 15n = 1,350 - 150$
$15n = 1,200$
$\frac{15n}{15} = \frac{1,200}{15}$
$n = 80$

The monthly fee is $80.

Make sure you are careful not to make any arithmetic errors!

NOTICE: Photocopying any part of this book is prohibited by law.

### Second Solution

**2. Select a Strategy**

Second, we'll review what Howard did using a strategy called **Working Backward**.

**3. Solve**

Second solution—**Work Backward.**

total − registration fee = total monthly fees
$1,350 − $150 = $1,200
total monthly fees/months = monthly fee
$1,200 ÷ $12 = $100
monthly fee = $100

### Tip

- Many problems can be solved using different strategies. As long as your choice leads to a correct answer and a correct explanation, it is a good choice!

# 2. What Is a Rubric?

## First Solution

### 4. Write/Explain

You must give a written explanation of how you solved the problem and what you were thinking. Clearly explain what you did and why you did it.

Do not leave out any steps.

> I wrote an Algebraic Equation. I let n represent the monthly fee. Since the registration fee of $150 plus the monthly fees for 15 months (15n) equals the total amount that Janine pays in 15 months, I wrote this Algebraic Equation: $150 + 15n = $1,350. When I solved the equation for n, I got n = $80. So, the monthly fee is $80.

### 5. Reflect

Janine reviewed her work by checking it against the rubric. She answered the problem that was asked, knew the facts, chose a good strategy, and used that strategy correctly. Janine's algebraic equation accurately represented the situation, and she took the correct steps to solve the equation. Her calculations were done correctly. Finally, Janine wrote a complete and clear answer to explain her solution.

### Score

Janine gets a score of **4**.

## Second Solution

### 4. Write/Explain

> I started with the total that Roberta has paid so far. That's $1,350. I subtracted the registration fee of $150 from the total. That left the amount Roberta has paid in monthly fees, which was $1,200. I divided that amount by the number of months Roberta has been a member, which is 12. That gave me a monthly fee of $100.

### 5. Reflect

Howard reviewed his work by checking it against the rubric. He answered the question that was asked. Howard chose a good strategy and he took the steps necessary in order to carry out the strategy. However, he made an arithmetic mistake in one of steps. Howard divided the total monthly fees by 12 months instead of 15 months. As a result, he got an incorrect answer.

### Score

Howard gets a score of **3**.

He could improve his work to get a **4** by dividing the total monthly fees by 15 instead of 12 and getting $80 as the monthly fee.

## 2. What Is a Rubric?

### Using the Rubric

Whenever you solve an open-ended math question, you should use the rubric on *page 13* as a guide.

Use the list in the **score of 4** box as a checklist. This will remind you what to include in your answer for the highest score possible.

### Reviewing Your and Your Partner's Work

After you finish solving the question, **self-assess**. This means that you should use the whole rubric to review your work and to score it. How well did you do? If you need to raise your score, take the time to do so.

You may also **peer-assess**. Swap your work with a partner. Use the rubric to score each other's solution. Now talk about the different strategies, answers, and the scores that were given. What might seem clear to you may not be clear to your friend. Partners can help each other learn what should be improved. You can discuss different ways to solve the same problem. The more you talk about your mathematics, the more you will understand how to improve your work.

### 2-Step Decision-Making Process

Some students find it easier before scoring with a rubric, to first use the **2-Step Decision-Making Process** as seen in the next column. It helps decide if you or your partner's answer is a **3 or 4** or a **1 or 2**.

NOTICE: Photocopying any part of this book is prohibited by law.

### 2-Step Decision-Making Process

**Before** you use a rubric, use a **2-Step Decision-Making Process**. This will give you a jump on the scoring your work or your partner's work.

Decide if your work is:

- *acceptable* (3 or 4)

or

- *unacceptable* (1 or 2).

If your work is *acceptable*, decide if it is:

- **full and complete (4)**

or

- **nearly full and complete, but not perfect (3).**

If your work is *unacceptable*, decide if it shows:

- **limited or only some understanding (2)**

or

- **little or no understanding (1).**

A **(0)** is **no attempt**.

Use your rubric to guide you as you solve problems. Check your answers against the qualities that the rubric lists for a perfect **4**. As you do this, you will see that your answers become much more clear and complete. Writing this kind of solution may become so much of a habit for you that you no longer will need the rubric to guide you. Then, you will only need the rubric to score your answer.

Mathematics Open-Ended Questions, Level G

# 3. How to Answer an Open-Ended Math Question

We've learned what an **open-ended math problem** is, how to solve it and how we can use a rubric to score it. Now we will analyze a problem and work through the solution together. Then we'll see how some other students solved the problem. We will use the **rubric** to score their solutions. Then we will discuss how these students could **improve** their answers and raise their scores.

NOTICE: Photocopying any part of this book is prohibited by law.

# 3. How to Answer...

## Modeled Problem

A **circular** garden has a **radius** of 5 feet. The garden is surrounded by a slate **border** that has a **width** of 2 feet at all points around the garden. What is the **area** of the slate border?

Keywords: circular, radius, border, width, area

### 1. Read and Think

What **question** are we asked?

- What is the area of the slate border?

What are the **keywords**?

- circular, radius, border, width, area

What **facts** are we given?

- A circular garden has a radius of 5 feet. The garden is surrounded by a slate border that has a width of 2 feet.

What is **going on**?

- There is a circular garden. There is a slate border that surrounds the garden. Since the border has the same width at all points around the garden, its outer edge also forms a circle.

### 2. Select a Strategy

We can **Draw a Picture** to help us visualize the situation. Drawing a picture is one of the strategies from the strategy list in Chapter One.

The garden is a circle with a radius of 5 feet. The garden is surrounded by a slate border that is 2 feet wide.

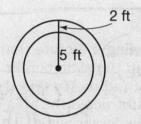

### 3. Solve

The diagram shows that the circular garden and the slate border, when combined, form a circle with a radius of 7 feet. It also indicates that to find the area of the slate border, we can subtract the area of the garden from the area of the larger circle.

Radius ($r$) of circular garden = 5 ft

Radius ($r$) of larger circle = 5 ft + 2 ft = 7 ft

Area ($A$) of circular garden = $\pi r^2$ = $3.14 \times (5 \text{ ft})^2 = 3.14 \times 25 \text{ ft}^2 = 78.5 \text{ ft}^2$

NOTICE: Photocopying any part of this book is prohibited by law.

# 3. How to Answer...

Area (A) of larger circle = $\pi r^2$ =
$3.14 \times (7 \text{ ft})^2 = 3.14 \times 49 \text{ ft}^2 = 153.86 \text{ ft}^2$

Area (A) of larger circle − Area (A) of circular garden = Area of slate border:

$153.86 \text{ ft}^2 - 78.5 \text{ ft}^2 = 75.36 \text{ ft}^2$

So, the area of the slate border is 75.36 ft².

## 4. Write/Explain

We started by **Drawing a Picture** that showed a circular garden and the slate border that made a circle with a 7-foot radius. To find the area of the slate border, we subtracted the area of the garden from the area of the larger circle. We found the difference, which was 75.36 ft². The area of the slate border is 75.36 ft².

## 5. Reflect

We reviewed our drawing, our thinking process, and our calculations. The drawing accurately represents the situation. Our method for finding the area of the border makes sense, given the information in the drawing. Each step of our process is clearly explained. Our calculations are correct.

### Score

This solution would earn a perfect **4** on our rubric.

- We showed that we knew what was asked and what information was given, including keywords and facts.
- We chose a good strategy and applied it correctly.
- We used the correct formulas, calculated correctly, and got the correct answer.
- We labeled our work correctly.
- We wrote a good explanation of how we used our strategy.
- We included all of our steps.
- We clearly explained what we did.
- We clearly explained our reasoning.

Now let's look at some answers that were written by other students.

# 3. How to Answer...

## Lana's Paper

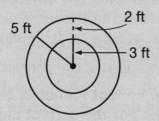

**Solve:** Area of garden = $\pi r^2$
= 3.14 × (3 ft)$^2$
= 28.26 ft$^2$
Area of garden + border = $\pi r^2$
= 3.14 × (5 ft)$^2$
= 3.14 × 25 ft$^2$ = 78.5 ft$^2$
78.5 ft$^2$ − 28.26 ft$^2$ = 50.24 ft$^2$
The area of the border is 50.24 ft$^2$.

**Write/Explain:** I drew a circle to show a garden and border that had a radius of 5 feet. Since the border was 2 feet of the 5 feet, I drew a smaller circler inside with a radius of 3 feet. That small circle was the garden. To find the area of the border, I subtracted the area of the small circle from the area of the large circle. I got an area of 50.24 ft$^2$.

Let's use our rubric to see how well Lana did.

- Did she show that she knew what the problem asked? **Yes, but she did not write the question.**
- Did she know what the keywords were? **Yes.**
- Did she show what facts were given? **Yes.**
- Did she name and use the correct strategy? Did she explain why and how she used the strategy? **She Drew a Picture, but the drawing was not correct.**
- Was her math correct? **No.**
- Was her answer correct? **No. Although Lana's computations were correct, they were not the correct solution to the problem. She got an area of 50.24 ft$^2$ instead of an area of 75.36 ft$^2$.**
- Did she label her work? **She needed to relabel parts of her drawing.**
- Were all of her steps included? **Yes.**
- Did she write a clear explanation of her work? **Yes.**

### Score

Lana would get a **3** on our rubric.

# 3. How to Answer . . .

She did not use the strategy correctly. Lana's drawing showed the 2-foot-wide circular border as part of the circle that had a radius of 5 feet. She should have showed the circular border so that it surrounded the circle that had the 5-foot radius. If she had done that, she would have found the correct area of the border. **To get a 4,** Lana should have reread the problem and compared it to her drawing. She would have realized that she needed to relabel the radii in her drawing.

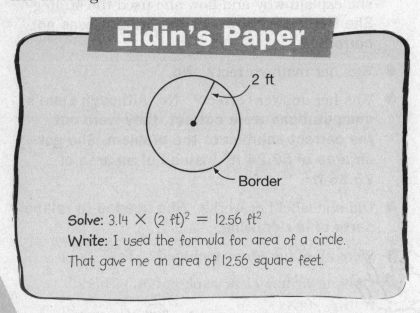

### Eldin's Paper

Solve: $3.14 \times (2 \text{ ft})^2 = 12.56 \text{ ft}^2$
Write: I used the formula for area of a circle. That gave me an area of 12.56 square feet.

- Did he show that he knew what the problem asked for?  **Yes, but he did not write the question.**
  - Did he know what the keywords were?  **No.**

- Did he show that he knew what facts were given?  **No. He didn't understand that the border surrounded a circular garden. He thought that the border was a circle, and that the width was its radius.**

- Did he name and use the correct strategy? Did he explain why and how he used the strategy?  **No. Although drawing a picture could work, Eldin's picture did not represent the situation.**

- Was his math correct?  **Yes, for the wrong problem.**

- Was his answer correct?  **No. Although Eldin correctly computed the area of a circle, his answer did not describe what was asked in the problem.**

- Did he label his work?  **No.**

- Were all of his steps included?  **No.**

- Did he write a good, clear explanation of his work?  **No. His explanation was incomplete.**

### Score

Eldin would receive a **2** on our rubric.

# 3. How to Answer . . .

Eldin understood that the problem asked for the area of the border, and that he needed to use the formula for the area of a circle. His calculations were correct. But he found the area of a circle with a radius of 2, instead of the border described in the problem. **To get a 4,** Eldin needs to understand that the problem describes a circular garden within a circular border. He needs to use this information to find the area of the border.

Janie knew that the problem involved a circle. But she did not understand the problem. She was not clear about the keywords or what they meant. She found the circumference of the garden instead of the area of the border. In addition, Janie did not write an explanation of her work. **To get a 4,** Janie needs to reread the problem and figure out what it is asking for. Then she needs to choose a good strategy and find another solution.

### Janie's Paper

Solve: $2\pi r = 2 \times 3.14 \times 5 \text{ ft} = 31.4 \text{ ft}^2$
Write: The area is 31.4 square feet.

### Score

Janie would receive a **1** on our rubric.

Mathematics Open-Ended Questions, Level G

# 4. How NOT to Get a Zero!

No one wants to get a **0** on an open-ended math problem. However, you can almost always get some points. The only person who gets a **0** is the person who **leaves the paper blank** or who **writes something that has nothing to do with the problem.** Let's see how we can start by scoring a **1 or 2** on our work, and then bring it up to a **3 or 4**.

And remember:

1. **Read and Think**
2. **Select a Strategy**
3. **Solve**
4. **Write/Explain**
5. **Reflect**

# 4. How NOT to Get a Zero!

## How to Get a 1 or 2!

Here is how to get **some credit** on an open-ended math question.

1. **Read** the question. Then **reread** it.

   **Ask:** "What are the **keywords** to help you solve the problem?"

   Finish the sentence: "**The keyword/s are**

   _____."

   **You will get credit for listing the keywords.**

2. **Understand** the problem. **Repeat the story** of the problem in your own words.

   **Ask:** "What am I **being asked** to do? What do I **need to find**?"

   Finish the sentence: "**I need to find**

   _____."

   **You will get credit for listing what was asked.**

3. **Find the facts** in the problem.

   **Ask:** "What does the problem tell me? What do I **know**?"

   Finish the sentence: "**The things I know are**

   _____."

   **You will get credit for writing the facts.**

4. Figure out what **strategy** you will use to help you solve the problem.

   **Ask:** "What **can help** me to find what I need to know?"

   Finish the sentence: "**The strategy I will use is**

   _____."

   **You will get credit for listing the strategy you use.**

## How to Get a 1 or 2!

You will practice these steps as you help solve the modeled problem introduced on the following page.

(continued on the next page.)

### Tip

- To get points right away, always begin by writing down what you are asked to find and what facts you are given.

NOTICE: Photocopying any part of this book is prohibited by law.

# 4. How NOT to Get a Zero!

Now let's work together on an open-ended math problem. First, let's try to get a **1 or 2**.

## Modeled Problem

Ida has 64 pink carnations and 40 yellow carnations. She wants to use the carnations to make bouquets that satisfy these conditions:

- **Each** bouquet is of **equal** size and has **both** colors in it.

- Each bouquet will have an equal number of pink carnations, and each bouquet will have an equal number of yellow carnations.

- There will be as many bouquets as possible.

How **many** bouquets will there be in all? How many pink carnations will be in each bouquet? How many yellow carnations will be in each bouquet?

**Keywords: each, equal, both, many**

### 1. Read and Think

1. Carefully **read** the question. **Reread** the question to fully understand it.

2. What are we **being asked**?
   - We need to find how many bouquets there will be in all. We also need to find how many pink carnations and how many yellow carnations will be in each bouquet.

   **By writing what you are being asked, you can get a score of a 1 or 2.**

3. Are there any **keywords** that can help you solve the problem? What are they?
   - each, both, many, equal

   **By listing keywords, you can get a score of a 1 or 2.**

4. What are the **facts**?
   - There are 64 pink carnations and 40 yellow carnations.
   - The carnations will be used to make bouquets of equal size.
   - Each bouquet will have both colors.
   - Each bouquet will have an equal number of pink carnations.
   - Each bouquet will have an equal number of yellow carnations.
   - There will be as many bouquets as possible.

   **By listing the facts, you can get a score of a 1 or 2.**

> **Hint**
> To get points right away, always begin by writing down **what you are asked** and what you are given as **facts**.

NOTICE: Photocopying any part of this book is prohibited by law.

# 4. How NOT to Get a Zero!

## 2. Select a Strategy

First, you have to pick a strategy and solve the problem. There are lots of strategies from which to pick. You may choose one from *pages 8-9* or use a different strategy of your own. For most problems, there is more than one strategy that will work. Sometimes, you will use a combination of strategies. The important thing is to write an answer that is clear, complete, and correct. You and a friend may use strategies that don't seem alike, and still get the same correct answer. Both of you can get a **4**.

1. What **strategy** will we use?

- **I will use a strategy called Divide and Conquer. First I will use Logical Thinking to analyze this problem. Then I will use the Make an Organized List strategy to help me solve the problem.**

**By writing what strategy you chose, you can get a score of a 1 or 2.**

**Hint:** We're going to use a lot of strategies in this book.

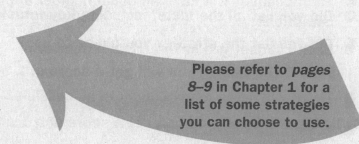

Please refer to *pages 8–9* in Chapter 1 for a list of some strategies you can choose to use.

## How to Get a 3 or 4!

*Now*, let's try to increase our score on the same problem from a **1 or 2** to a **3 or 4**!

First, let's review. *Remember, you can always get some credit* for listing the keywords and the question that is asked. You will also receive points by writing the facts that are given. Finally, credit will be given for listing the strategy you have chosen. By doing this, you will receive a score of at least a **1 or 2**. Now it is time to raise your score to a **3 or 4**. Use all the following information to do so.

So here we go:
Here is how to take a score of **1 or 2** and make it a **3 or 4.**
We will continue using the same modeled problem.

**Hint:** An open-ended math problem can be solved in more than one way. And if your way works out and gives the correct answer, then you are right!

### 3. Solve

Go back to the problem. Analyze the facts and conditions that the problem describes:

- There are 64 pink carnations and 40 yellow carnations.
- The carnations will be used to make bouquets of equal size.
- Each bouquet will have both colors.
- Each bouquet will have an equal number of pink carnations.

NOTICE: Photocopying any part of this book is prohibited by law.

# 4. How NOT to Get a Zero!

- **Each bouquet will have an equal number of yellow carnations.**

- **There will be as many bouquets as possible.**

The pink carnations and yellow carnations will be put together in a certain number of bouquets. This means they will each be separated into the same number of groups. To do this we will use the **Divide and Conquer** strategy. *First,* to find the greatest number of groups, find the greatest common factor (GCF) of 64 and 40. One way to do this is to **Make an Organized List** that shows the factors of each number.

Factors of 64: 1, 2, 4, **8**, 16, 32, 64

Factors of 40: 1, 2, 4, 5, **8**, 10, 20, 40

The organized list shows that the GCF of 64 and 40 is 8. So, the greatest number of bouquets that can be made is 8. *Now*, we will **Write a Number Sentence** in which we divide to find how many carnations of each color will be in a bouquet.

$64 \div 8 = 8$  There will be 8 pink carnations in each bouquet.

$40 \div 8 = 5$  There will be 5 yellow carnations in each bouquet.

## 4. Write/Explain

The person marking your paper does not know what you were thinking. You must explain why you chose the strategy, and how you solved the problem. Be sure to label your work. Don't leave out any steps. Reread your writing, making sure your work is clear and complete.

So here we go:

In order to solve this problem we used the **Divide and Conquer** strategy. We needed to separate 64 pink carnations and 40 yellow carnations into equal groups. By **Making an Organized List**, we found the Greatest Common Factor of 64 and 40, which was 8, the greatest number of possible bouquets. Then we **Wrote a Number Sentence**. We divided the numbers of flowers of each color by the number of bouquets. That gave us 8 pink carnations and 5 yellow carnations in each bouquet.

## 5. Reflect

**Now Review Your Work and Improve It!**

After you solve the problem, carefully review your work.

- **Did you write what the problem asked you to find?**

- **Did you list all the facts, including keywords?**

- **Did you list the strategy you chose to use?**

**If you did these things, you will get a score of 1 or 2**

- **Did you use the right strategy?**

- **Is your arithmetic right?**

NOTICE: Photocopying any part of this book is prohibited by law.

# 4. How NOT to Get a Zero!

- Did you label your work?
- Did you write out all the steps to solving the problem?
- Did you explain why you chose the strategy and how you used it?
- Did you explain why you solved the problem the way you did?
- Is your writing clear?

**If you did these things, you will raise your score from a 1 or 2 to a 3 or 4.**

If you do all the things we have suggested, you **CANNOT** get a **0**.

**Remember, never leave your paper blank.**

## Work with Peers

You and a classmate can exchange your papers. Your classmate can tell you if he or she understood your solution process and your explanation. That will help you see how clearly you explained your work. If you get into the habit of writing clear explanations, that's a good way to help yourself get a **3** or a **4**.

### Use a Checklist and Your Rubric

In the next column is a checklist that will help you make sure you have done the best that you can. You can use this checklist and the rubric that was in Chapter Two to improve your work.

## Read It and Think

- ☐ Did I read the problem at least twice? Do I understand it?
- ☐ Did I write down the question being asked?
- ☐ Did I write down the keywords in the problem?
- ☐ Did I write down the facts that are given?
- ☐ Did I write down the strategy that I used?
- ☐ Did I solve the problem?
- ☐ Is my arithmetic correct? Are my calculations correct?
- ☐ Did I explain how I solved the problem?
- ☐ Did I explain why I chose the strategy and how I used it?
- ☐ Did I include all the steps I took to solve it?
- ☐ Is my writing clear?
- ☐ Did I label my work?
- ☐ Does my answer make sense?
- ☐ Did I answer the exact question being asked?

**Always check your work!**

**Mathematics Open-Ended Questions, Level G**

# 5. Number and Operations

There are **lots of different kinds of numbers and lots of different ways to use them**. We can use fractions and mixed numbers like $\frac{1}{2}$ and $3\frac{3}{8}$ to measure things, such as cups of flour or miles traveled. We can use numbers written in scientific notation such as $4.9 \times 10^{-9}$ or $1.4 \times 10^{23}$ to represent very small or very large amounts, such as the diameter of a hydrogen atom in centimeters, or the distance from Earth to a quasar in miles. We can use different ways to express the same amount such as $\frac{1}{4}$, 0.25, or 25%. In this chapter we will work with all kinds of numbers. We will add, subtract, multiply, divide, and perform combinations of operations. We will use our understanding of the meanings of numbers to help us solve problems.

# 5. Number and Operations

Here is a problem that you might have to solve on a test. Let's solve it together to show what a model **4** answer might look like. Then we can score it using a **rubric**.

## Modeled Problem

A taxicab charges $1.60 for the **first** mile and $0.80 for each **additional** mile. Marie spent $15 on her cab ride, which **includes** a $2.20 tip. How **many** miles long was Marie's cab ride?

Keywords: first, additional, includes, many

### 1. Read and Think

Let's **read** the problem.

What **question** are we asked?

- We are asked to find how many miles long Marie's cab ride was.

Do we recognize what the **keywords** are?

- first, additional, includes, many

What **facts** are we given?

- The taxicab charges $1.60 for the first mile.
- The taxicab charges $0.80 for each additional mile.
- Marie spent $15, which included a $2.20 tip.

### 2. Select a Strategy

We know the total that Marie spends on the cab ride. We also know the rate for each mile and the tip that she gives. So, we will use the **Work Backward** strategy to find how many miles long the cab ride is. We will start with the total and work backward to find the answer.

### 3. Solve

First, let's find what the cab charged for the miles that Marie traveled. To do this, we need to subtract the tip.

$$\$15.00 - \$2.20 = \$12.80$$

Next, let's subtract the cost of the first mile from the cost without the tip. That will give us the amount Marie spent on the additional miles after the first mile.

$$\$12.80 - \$1.60 = \$11.20$$

NOTICE: Photocopying any part of this book is prohibited by law.

# 5. Number and Operations

Next, let's find the number of additional miles Marie traveled at $0.80 per mile.

Marie spent a total of $11.20 on the additional miles after the first mile. Since the rate was $0.80 per mile, we divide the $11.20 by $0.80 to find the number of additional miles.

$$\$11.20 \div \$0.80 = 14 \text{ miles}$$

### 4. Write/Explain

We subtracted the tip from the total Marie spent to give us $12.80 for the miles Marie traveled. Next, we subtracted the charge for the first mile from the total charge to give us $11.20 for the charge for the additional miles. Then, we divided that charge by the rate per additional mile to find the number of additional miles, which was 14. So, Marie traveled 15 miles.

### 5. Reflect

Let's review our work and answer.

- Did we show that we knew what the problem asked? **Yes.**
- Did we know what the keywords were? **Yes.**
- Did we show that we knew what facts were given? **Yes.**
- Did we name and use the correct strategy? **Yes.**
- Was our math correct? **Yes.**
- Did we label our work? **Yes.**
- Was our answer correct? **Yes.**
- Did we explain why we chose the strategy and how it was used to get the right answer? **Yes.**
- Were all of our steps included? **Yes.**
- Did we write a good, clear explanation of our work? **We reread the problem and decided that it was clear and complete. A friend would understand it.**

### Score

This solution would earn a **4** on our rubric. The answer is correct and the solution process is clearly explained.

# 5. Number and Operations

Here are some **Guided Open-Ended Math Problems.** For each problem there are **four parts.** In the **first part,** you will solve the problem with guided help. In the **second part,** you will score and correct a solution with guided help. The **third part** shows one solution that scores a perfect **4**. This solution may or may not differ from your way. The **fourth part** has *answers* to the **first** and **second parts** so you can check your work.

## Guided Problem #1

Mr. Taylor withdrew $220 from the bank. The bank teller gave Mr. Taylor 16 bills, which were all $10 bills and $20 bills. How many $10 bills and how many $20 bills did the bank teller give Mr. Taylor?

Keywords: ? ?

### 1. Try It Yourself.

Answer the questions below to get a score of **4**.

What **question** are you being asked?

_____

What are the **keywords**?

What are the **facts** you need to solve the problem?

_____
_____
_____

What **strategy** can you use to solve the problem?

_____
_____

**Solve** the problem.

_____
_____
_____

**Hint**
Possible answers include **Guess and Test** and **Make an Organized List or Table.**

**Write/Explain** what you did to solve the problem.

_____
_____
_____

**Reflect.** Review and improve your work.

_____
_____

Use the rubric on *page 13* to score this work.

NOTICE: Photocopying any part of this book is prohibited by law.

33

# 5. Number and Operations

### 2. Donna Tries It.

## Donna's Paper

**Question:** How many $10 bills and $20 bills did the bank teller give Mr. Taylor?

**Keywords:** withdrew, many, $10 bill, $20 bill

**Facts:** There are 16 bills worth a total of $220.
The bills are all $10 bills and $20 bills.

**Strategy:** I will use the Guess and Test strategy.

**Solve:** Guess #1: 7 tens and 9 twenties = 16 bills
(7 × $10) + (9 × $20) =
$70 + $180 = $250
The bills are worth a total of $250.
$250 > $220.
To lower the total to $220, I try a combination with fewer twenties and more tens.
Guess #2: 12 tens and 4 twenties = 16 bills
(12 × $10) + (4 × $20) =
$120 + $100 = $220
Mr. Taylor got 12 tens and 4 twenties from the teller.

**Write/Explain:** I used the Guess and Test strategy. My first guess of 7 tens and 9 twenties gave me $250. So, I changed the number of tens and twenties so that the total value of the money would be smaller. I found that 12 tens and 4 twenties equal $220.

**Score the Answer.**

According to the rubric, from **1** to **3** what score would you give Donna? Explain why you gave that score.

_____
_____
_____
_____
_____
_____
_____

**Make it a 4!** Rewrite.

_____
_____
_____
_____
_____
_____
_____
_____

Use the rubric on *page 13* to score her problem.

NOTICE: Photocopying any part of this book is prohibited by law.

# 5. Number and Operations

## 3. Hector Tries It.

Remember there is often more than one way to solve a problem. Here is how Hector solved this problem.

### Hector's Paper

**Question:** How many $10 bills and how many $20 bills did the teller give to Mr. Taylor?
**Keywords:** withdrew, $10 bill, $20 bill, many
**Facts:** total of 16 bills
  bills are all $10 and $20
  bills are worth $220
**Strategy:** I decided to Make a Table.
**Solve:** The table showed possible combinations of 16 $10 bills and $20 bills.

| $10 bills | $20 bills | Total Value |
|---|---|---|
| 8 | 8 | (8 × $10) + (8 × $20) = $80 + $160 = $240 |
| 9 | 7 | (9 × $10) + (7 × $20) = $90 + $140 = $230 |
| 10 | 6 | (10 × $10) + (6 × $20) = $100 + $120 = $220 |

The bank gave Mr. Taylor 10 $10 bills and 6 $20 bills.
**Write/Explain:** I Made a Table to list combinations of $10 bills and $20 bills in which there were 16 bills. I started with 8 $10 bills and 8 $20 bills, which gave me a total of $240. $240 is greater than $220.50. I decreased the number of $20 bills by one and increased the number of $10 bills by one until I found a combination with a value of $220. That combination was 10 $10 bills and 6 $20 bills.

**Score:** Hector's solution would earn a 4 on a test. Hector identified the question that was asked, and the keywords and facts. He picked a good strategy, used it correctly, and explained why and how he used it. He clearly explained the steps taken to solve the problem and labeled his work. It is perfect!

NOTICE: Photocopying any part of this book is prohibited by law.

# 5. Number and Operations

## 4. Answers to Parts 1 and 2.

### Guided Problem #1

Mr. Taylor withdrew $220 from the bank. The bank teller gave Mr. Taylor 16 bills, which were all $10 bills and $20 bills. How many $10 bills and how many $20 bills did the bank teller give Mr. Taylor?

Keywords:

**1. Try It Yourself. (page 33)**

**Question:** How many $10 bills and how many $20 bills did the bank teller give Mr. Taylor?

**Keywords:** withdrew, $10 bill, $20 bill, many

**Facts:** Mr. Taylor withdrew $220 from the bank.
Mr. Taylor got a total of 16 bills.
The bills are all $10 bills and $20 bills.

**Strategy:** Guess and Test

**Solve:** I used the Guess and Test strategy.

Guess #1: 8 $10 bills and 8 $20 bills

$(8 \times \$10) + (8 \times \$20) = \$80 + \$160 = \$240$

Guess #2: 9 $10 bills and 7 $20 bills

$(9 \times \$10) + (7 \times \$20) = \$90 + \$140 = \$230$

Guess #3: 10 $10 bills and 6 $20 bills

$(10 \times \$10) + (6 \times \$20) = \$100 + \$120 = \$220$

The bank teller gave 10 $10 bills and 6 $20 bills to Mr. Taylor.

**Write/Explain:** I used the **Guess and Test** strategy to look for a combination of 16 $10 bills and $20 bills that had a value of $220. I tried different combinations until I found a group that was worth $220, which was 10 $10 bills and 6 $20 bills.

**2. Donna Tries It. (page 34)**

**Score the Answer:** I would give Donna a **3**. Donna knew what the problem asked. She listed the keywords and knew what facts were given. She labeled her work. Donna used the Guess and Test strategy correctly and explained why and how she used it. But on the second guess, she made an error when calculating that 12 $10 bills and 4 $20 bills make $2000.

**Make it a 4!** Rewrite.

Correct the value of Donna's second guess and use it to find the solution.

Guess #2: 12 tens and 4 twenties = 16 bills

$(12 \times \$10) + (4 \times \$20) + = \$120 + \$80 = \$200$

$200 < 220$. To make $220, I will try a combination with more twenties and fewer tens.

Guess #3: 10 tens and 6 twenties = 16 bills

$(10 \times \$10) + (6 \times \$20) = \$100 + \$120 = \$220$

NOTICE: Photocopying any part of this book is prohibited by law.

# 5. Number and Operations

The bank teller gave 10 $10 bills and 6 $20 bills to Mr. Taylor.

## Guided Problem #2

The Sports and Outdoor Shop has an opening-day promotion. Every 12th person who makes a purchase gets a free backpack, and every 15th person who makes a purchase gets a free tent. During opening day, 185 people make purchases. How many people received both a free backpack and a free tent?

Keywords: ??

### 1. Try It Yourself.

Answer the questions below to get a score of **4**.

What **question** are you being asked?

_____

_____

What are the **keywords**?

_____

_____

What are the **facts** you need to solve the problem?

_____

_____

_____

_____

What **strategy** can you use to solve the problem?

_____

_____

**Solve** the problem.

_____

_____

_____

_____

**Hint**

Possible answers include **Divide and Conquer, Logical Thinking,** and **Make an Organized List.**

**Write/Explain** what you did to solve the problem.

_____

_____

_____

**Reflect.** Review and improve your work.

_____

_____

Use the rubric on *page 13* to score this work.

NOTICE: Photocopying any part of this book is prohibited by law.

# 5. Number and Operations

### 2. Teddy Tries It.

Use the rubric on *page 13* to score his problem.

#### Teddy's Paper

**Question:** How many people received a free backpack and a free tent?

**Keywords:** 12th, 15th, many, both

**Facts:** Every 12th person gets a free backpack. Every 15th person gets a free tent.

**Solve:**

```
        15 R5           12 R5
     12)185           15)185
       -120             -150
        ───              ───
         65               35
        -60              -30
        ───              ───
          5                5
```

15 people get a free backpack and 12 people get a free tent.

**Write/Explain:** I divided the number of customers by 12 to find how many people received free backpacks. Then I divided the number of customers by 15 to find how many customers received free tents. Fifteen customers received free backpacks and 12 customers received free tents.

### Score the Answer.

According to the rubric, from **1** to **3** what score would you give Teddy? Explain why you gave that score.

_____
_____
_____
_____
_____
_____

**Make it a 4!** Rewrite.

_____
_____
_____
_____
_____
_____
_____
_____

Use the rubric on *page 13* to score his problem.

NOTICE: Photocopying any part of this book is prohibited by law.

# 5. Number and Operations

## 3. Sabrina Tries It.

Remember there is often more than one way to solve a problem. Here is how Sabrina solved this problem.

### Sabrina's Paper

**Question:** How many people get both a free backpack and a free tent?

**Keywords:** 12th, 15th, many, both

**Facts:** Every 12th person who makes a purchase gets a free backpack. Every 15th person who makes a purchase gets a free tent. 185 people make purchases.

**Strategies:** Logical Thinking and the Divide and Conquer strategy.

**Solve:** I first use Logical Thinking. I want to find out how many people get both a free backpack and a free tent. Since every 12th person gets a free backpack, I can use multiples of 12 to find which people get a free backpack. Since every 15th person gets a free tent, I can use multiples of 15 to find which people get a free backpack. So, the first person that will get both a free backpack and free tent will be the Least Common Multiple (LCM) of 12 and 15.

I can use prime factorization to find the LCM of 12 and 15.

$$12 = 2 \times 2 \times 3$$
$$15 = \downarrow \quad \downarrow \quad 3 \times 5$$
$$\downarrow \quad \downarrow \quad \downarrow \quad \downarrow$$
$$LCM = 2 \times 2 \times 3 \times 5 = 60$$

Since the LCM of 12 and 15 is 60, that means that the 60th person is the first to get both a free backpack and a free tent. It also means that every 60th person gets both a free backpack and a free tent.

Now, I use the Divide and Conquer strategy. To find how many people will get both a free backpack and a free tent, I find the multiples of 60 are less than 185: 60, 120, 180. That means that 60th person will get both a free backpack and a free tent, and so will the 120th and the 180th person. That makes a total of three people out of the 185 people that will get a free backpack and a free tent.

**Write/Explain:** I used Logical Thinking and the Divide and Conquer strategy to solve the problem. Since each multiple of 12 represents a free backpack and each multiple of 15 represents a free tent, I found the LCM of the two numbers, which is 60. That means that every 60th person got both a free backpack and a free tent. So, to find how many people out of the 185 got both, I had to find how many multiples of 60 were less than 185. There were 3. That meant that 3 out of the 185 people got both a free backpack and a free tent.

(cont.)

NOTICE: Photocopying any part of this book is prohibited by law.

# 5. Number and Operations

(Sabrina's paper cont.)

**Score:** Sabrina's solution would earn a **4** on our rubric. She identified the question that was asked, the keywords, picked a good combination of strategies and knew the facts. She clearly explained the steps taken to solve the problem and why and how she used the strategy she did. She got the correct answer and labeled it.

### 4. Answers to Parts 1 and 2.

## Guided Problem #2

The Sports and Outdoor Shop has an opening-day promotion. Every 12th person who makes a purchase gets a free backpack, and every 15th person who makes a purchase gets a free tent. During opening day, 185 people make purchases. How many people received both a free backpack and a free tent?

**Keywords:** ? ?

### 1. Try It Yourself. (page 37)

**Question:** How many people get both a free backpack and a free tent?

**Keywords:** 12th, 15th, many, both

**Facts:** Every 12th person who makes a purchase gets a free backpack.

Every 15th person who makes a purchase gets a free tent.

185 people make purchases.

**Strategy:** Make an Organized List

**Solve:** To determine which people get both a free backpack and a free tent, I found the common multiples of 12 and 15 that are less than 185.

Multiples of 12: 12, 24, 36, 48, **60,** 72, 84, 96, 108, **120,** 132, 144, 156, 168, **180**

Multiples of 15: 15, 30, 45, **60,** 75, 90, 105, **120,** 135, 150, 165, **180**

Three people get both a free backpack and a free tent.

**Write/Explain:** I used the **Make an Organized List** strategy. Since every 12th person got a free backpack and every 15th person got a free tent, I listed the multiples of 12 and 15 that were less than 185. My list showed 3 common multiples that were less than 185. So, 3 people got both a free backpack and a free tent.

# 5. Number and Operations

## 2. Teddy Tries It. (page 38)

**Score the Answer:** Teddy would get a **2**. He listed the question, the keywords, and the facts. He labeled his work. Teddy's calculations were correct. But they were not the calculations that were needed to answer the question. Teddy answered a question that was different from the question that was asked. He also didn't name the strategy that he used.

**Make it a 4!** Rewrite.

Find the common multiples of 12 and 15 that are less than 185.

Multiples of 12: 12, 24, 36, 48, **60,** 72, 84, 96, 108, **120,** 132, 144, 156, 168, **180**

Multiples of 15: 15, 30, 45, **60,** 75, 90, 105, **120,** 135, 150, 165, **180**

I used the **Make an Organized List** strategy. Since every 12th person got a free backpack and every 15th person got a free tent, I listed the multiples of 12 and 15 that were less than 185, which was the total number of people. My list showed 3 common multiples that were less than 185. So, 3 people got both a free backpack and a free tent.

## Guided Problem #3

Latoya has the discount coupons shown below.

| Mel's Music Mart | Mel's Music Mart |
|---|---|
| Buy 5 CDs Get the Sixth CD Free! | 20% off all CDs! |
| May not be used with any other coupon. | May not be used with any other coupon. |

Latoya wants to buy 6 CDs that cost $14 each. Which coupon should Latoya use to get the cheaper price? How much will she save by using that coupon?

Keywords:

### 1. Try It Yourself.

Answer the questions below to get a score of **4**.

What **question** are you being asked to find?

_____

_____

What are the **keywords**?

_____

_____

_____

NOTICE: Photocopying any part of this book is prohibited by law.

# 5. Number and Operations

What are the **facts** you need to solve the problem?

_____

_____

_____

What **strategy** can you use to solve the problem?

_____

_____

_____

**Hint**

Possible answers include **Logical Thinking**, and **Divide and Conquer**.

**Solve** the problem.

_____

_____

_____

**Write/Explain** what you did to solve the problem.

_____

_____

_____

_____

**Reflect.** Review and improve your work.

_____

_____

_____

_____

## 2. Raul Tries It.

### Raul's Paper

**Question:** Which coupon should Latoya use? How much will she save in all?

**Keywords:** discount, each, cheaper, save, much

**Facts:** Buy 5 CDs and get the sixth CD free with one coupon. Or, get 20% off all CDs with another coupon. CDs cost $14 each. Latoya wants to buy 6 CDs.

**Strategy:** Write Number Sentences

**Solve:** Get 1 CD free with 5 CDs: $\frac{1}{5} = 20\%$
The two coupons are both the same. They both give 20% off.

$6 \times \$14 = \$84$

$20\% \times \$84 = 0.2 \times 84 = \$16.80$

**Write/Explain:** The coupons were both for 20% off. I multiplied 6 CDs × $14 = $84. Then I multiplied to find 20% × $84 = $16.80. You would save $16.80 no matter which coupon you use.

Use the rubric on *page 13* **to score this work.**

NOTICE: Photocopying any part of this book is prohibited by law.

# 5. Number and Operations

**Score the Answer.**

According to the rubric, from **1** to **3** what score would you give Raul? Explain why you gave that score.

_____

_____

_____

**Make it a 4!** Rewrite.

_____

_____

_____

_____

_____

### 3. Barry Tries It.

Remember there is often more than one way to solve a problem. Here is how Barry solved this problem.

## Barry's Paper

**Question:** Which coupon should Latoya use? How much will she save in all?

**Keywords:** discount, each, cheaper, save, much

**Facts:** One coupon gives you a sixth CD free when you buy 5 CDs.
One coupon gives you 20% off all CDs. Latoya wants to buy 6 CDs that cost $14 apiece.

**Strategy:** Divide and Conquer

**Solve:** Coupon #1: Buy 5 at $14 apiece. Get the sixth free. You will save $14, the price of the sixth CD.
Coupon #2: Buy 6 at $14 apiece. Get 20% off.
Regular price: $6 \times \$14 = \$84$
Discount: $0.2 \times \$84 = \$16.80$
Latoya should use the coupon for 20% off. She will save a total of $16.80.

**Write/Explain:** I used the Divide and Conquer strategy to break the problem into parts. I found the amount that would be saved using each coupon. For the first coupon, the amount saved was the price of the sixth CD: $14. For the second coupon, I found the regular cost of the 6 CDs, multiplied the regular cost by the 20% discount, and found a savings of $16.80. Since $16.80 is more than $14, Latoya should use the coupon for 20% off.

**Score:** Barry would earn a 4 for his work. He identified the questions that were asked, as well as the keywords, and facts. He picked a good strategy, explaining why he chose it, and how he used it. He clearly explained the steps he used to solve the problem. Barry also labeled his work.

# 5. Number and Operations

## 4. Answers to Parts 1 and 2.

### Guided Problem #3

Latoya has the discount coupons shown below.

| Mel's Music Mart | Mel's Music Mart |
|---|---|
| Buy 5 CDs | 20% off all CDs! |
| Get the Sixth CD Free! | |
| May not be used with any other coupon. | May not be used with any other coupon. |

Latoya wants to buy 6 CDs that cost $14 each without a coupon. Which coupon should Latoya use to get the cheaper price? How much will she save by using that coupon?

Keywords:

## 1. Try It Yourself. (pages 41–42)

**Question:** Which coupon should Latoya use? How much will she save in all?

**Keywords:** discount, each, cheaper, save, much

**Facts:** Buy 5 CDs and get the 6th CD free with a coupon. Get 20% off all CDs with the other coupon. CDs cost $14 apiece. Latoya wants to buy 6 CDs.

**Strategy:** Logical Thinking

**Solve:** First coupon: 1 free out of $6 = \frac{1}{6} = 16\frac{2}{3}\%$

$16\frac{2}{3}\% < 20\%$. So, use the coupon for 20% off.

20% of regular cost of 6 CDs = 
$0.2(6 \times \$14) = 0.2 \times \$84 = \$16.80$

The amount saved is $16.80.

**Write/Explain:** I used **Logical Thinking** to choose which coupon to use. I figured out the percent discount offered by the first coupon and chose the second coupon because it offered a greater discount. Then I wrote a number sentence to figure out the amount saved. I multiplied the discount times the total cost of six CDs to get the amount saved.

NOTICE: Photocopying any part of this book is prohibited by law.

# 5. Number and Operations

## 2. Raul Tries It. (page 42)

**Score the Answer:** Raul would get a **2**. He listed the question, keywords, facts, and the strategy he used. He labeled his work. He multiplied correctly when finding the total regular cost and a 20% discount on that amount. He explained why and how he chose the strategy he did. But Raul did not correctly compare the discounts offered by the two coupons. Raul thought that they offered the same discount, when they did not.

**Make it a 4!** Rewrite!

Compare the rate of discount offered by each coupon and choose the coupon that offers the greater discount.

First coupon: 1 free out of 6 = $\frac{1}{6}$ = $16\frac{2}{3}$%

$16\frac{2}{3}$% < 20%, so, use the coupon for 20% off.

To find the amount saved using the 20% coupon, Write a Number Sentence.

20% of regular cost of 6 CDs =
0.2(6 × $14) = 0.2($84) = $16.80

The amount saved is $16.80.

I used **Logical Thinking** to choose which coupon to use. I figured out the percent discount offered by the first coupon, and used the second coupon because it offered a greater discount. Then I **Wrote a Number Sentence** to figure out the amount saved. I multiplied the discount times the total cost of six CDs to find the amount saved.

# 5. Number and Operations

# Quiz Problems

Here are some problems for you to try. Keep your **rubric** handy while you solve the problem. Let's see if you can score a **4**.

**1.** The Sands Hotel and The Breakers Hotel stand next to each other on the ocean. The Sands Hotel has 120 rooms, 72 of which are occupied. The Breakers Hotel has the same percent of rooms occupied as The Sands Hotel. The Breakers Hotel has 320 rooms. How many rooms in The Breakers Hotel are occupied?

**2.** At a blood bank, Jim donates one pint of blood. The nurse says to him, "Thanks, Jim! You have just donated $2.2 \times 10^{12}$ red blood cells!" If Jim's body generally has 11 pints of blood, about how many red blood cells are in his body? Does he have more than 10 trillion red blood cells?

**3.** The Rea family rents movies at Movie Station for $3.75 each for a 3-day period. They watch one movie each weekend and incur a late fee of $2.50 once a month. Movie Station also offers another payment plan of an unlimited number of rentals, but only one at a time, for $200 per year. Which payment plan offers the Rea family better value?

NOTICE: Photocopying any part of this book is prohibited by law.

# 5. Number and Operations

**4.** Nadia is building a decorative wall between two parts of a flower garden. She uses 6-inch-long bricks for the bottom row, 10-inch-long bricks for the second row, and 12-inch-long bricks for the third row. The wall is 20 feet long. Excluding the two ends of the wall, in how many places do all three kinds of bricks line up?

**5.** After a major snowstorm, Anthony and Stephen shoveled their neighbor's driveway in $1\frac{1}{2}$ hours. Dana joined them for the last 45 minutes. Their neighbor gave them $20, which they divided so that they each received the same rate of pay. Was their hourly rate of pay more or less than $5.15 per hour?

**6.** A football team needs 10 yards in 4 downs to gain a first down. The position of the ball changes by the same number of yards on the first down and on the second down. On the third down, the team gains 12 yards. On fourth down, the team still has 6 yards to go for a first down. What was the change in position of the ball on each of the first two downs?

**7.** The Stanford Middle School is presenting a production of "West Side Story." The school needs to gross $6,562.50 on seating to meet the play's expenses. If tickets are $12.50 each and the seating capacity is 750, what percent of capacity needs to be sold?

NOTICE: Photocopying any part of this book is prohibited by law.

# 6. Algebra

In this chapter, we will solve problems that involve algebra. In an algebra problem, we might use an **Algebraic Equation** such as $6n + 3 = 33$ to represent a situation that was described in words. We might look for a **Pattern or Rule** and use an **Algebraic Expression** or Equation to describe it. We might also **Use a Graph or Table** to show the relationship between two changing quantities, such as miles traveled and time, or age and height. These are just a few of the ways in which algebra might be used in problem-solving situations.

# 6. Algebra

In the modeled problem below, we will use algebra to describe a relationship between two amounts. Let's solve this problem together. Then we can use a **rubric** to score our answer. Let's try to get a score of **4**.

## Modeled Problem

The spreadsheet gives information about some items that are sold at the Maritime Gift Shop.

| A | B | C | D | E |
|---|---|---|---|---|
| Item | Unit Cost | Regular Price | Discount | Sale Price |
| Pocket Telescope | $30 | $52.50 | $10.50 | $42.00 |
| Porthole Barometer | $90 | $157.50 | $31.50 | $126.00 |
| Scale Model of *Yankee Clipper* ship | $120 | $210.00 | $42.00 | $168.00 |

What **equation** represents the relationship between the **regular price** and the **unit cost** of **each** item?

**Keywords: equation, regular price, unit cost, each**

NOTICE: Photocopying any part of this book is prohibited by law.

# 6. Algebra

### 1. Read and Think

Let's **read** the problem.

What **question** are we asked?

- **What equation represents the relationship between the regular price and the unit cost of each item?**

Do we recognize what the **keywords** are?

- **Yes. We listed the keywords.**

What **facts** are we given?

- **We know the unit price, regular price, discount, and sale price for three different items that are sold at the store.**

What **facts** do we need?

- **We only need the unit price and the regular price for each item.**

### 2. Select a Strategy

The problem asks us to relate the unit cost and the regular price for three items. To do this, we can take those two pieces of data for each item and use the **Look For a Pattern** strategy.

### 3. Solve

Let's compare the unit price and the regular selling price for each item and see if we can find a pattern. One way to compare the two amounts is to divide the regular selling price by the unit price for each item. By doing this, we can see if in each case the unit price was multiplied by the same factor to get the selling price.

Regular Price/Unit Cost

$$\$52.50/\$30 = 1.75$$

$$\$157.50/\$90 = 1.75$$

$$\$210/\$120 = 1.75$$

For each item, the regular price is 1.75 times the unit cost. If we let $C$ represent the regular price, and $B$ represent the unit cost, we can use the following equation to represent the relationship between those two amounts: $C = 1.75B$.

### 4. Write/Explain

To solve this problem, we compared the unit price and the regular price for each item and **Looked For a Pattern.** For each item, we divided the regular price by the unit cost and found that the regular price was 1.75 times the unit cost. So, $C = 1.75B$.

# 6. Algebra

## 5. Reflect

Let's review at our work and answer.

- Did we show that we knew what the problem asked?  **Yes. We Wrote an Equation that told how the regular selling price was related to the unit cost.**

- Did we know what the keywords were?  **Yes.**

- Did we show that we knew what facts were given?  **Yes.**

- Did we name and use the correct strategy?  **Yes. Looking For a Pattern allowed us to find out and describe how the regular price was related to the unit cost.**

- Was our math correct?  **Yes.**

- Did we label our work?  **Yes.**

- Was our answer correct?  **Yes.**

- Did we explain why we chose the strategy and how it was used to get the right answer?  **Yes.**

- Were all of our steps included?  **Yes.**

- Did we write a good, clear explanation of our work?  **We reread the problem and decided that it was clear and complete. A friend would understand it.**

### Score

This solution would earn a **4** on our rubric. The answer is correct and the solution process is clearly explained.

On the following pages are some **Guided Open-Ended Math Problems**. For each problem there are **four parts**. In the **first part**, you will solve the problem with guided help. In the **second part**, you will score and correct a solution with guided help. The **third part** shows one solution that scores a perfect **4**. This solution may or may not differ from your way. The **fourth part** has *answers* to the **first** and **second parts** so you can check your work.

*Use the rubric on page 13 to score this work.*

NOTICE: Photocopying any part of this book is prohibited by law.

# 6. Algebra

## Guided Problem #1

The 10 members of the Debating Club are in a room. Five are wearing green shirts. Five are wearing blue shirts. Each member shakes hands with each other exactly once. How many handshakes will there be altogether?

Keywords: **? ?**

### 1. Try It Yourself.

Answer the questions below to get a score of **4**.

What **question** are you being asked?

_____
_____
_____

What are the **keywords**?

_____
_____
_____

What are the **facts** you need to solve the problem?

_____
_____
_____

What **strategy** can you use to solve the problem?

_____
_____
_____
_____

**Solve** the problem.

_____
_____
_____
_____

> **Hint**
>
> Possible answers include: **Make it Simpler, Write a Number Sentence or an Algebraic Equation, Draw a Picture, Act it Out, Look for a Pattern,** and **Logical Thinking.**

**Write/Explain** what you did to solve the problem.

_____
_____
_____
_____

**Reflect.** Review and improve your work.

_____
_____
_____
_____

NOTICE: Photocopying any part of this book is prohibited by law.

# 6. Algebra

## 2. Diallo Tries It.

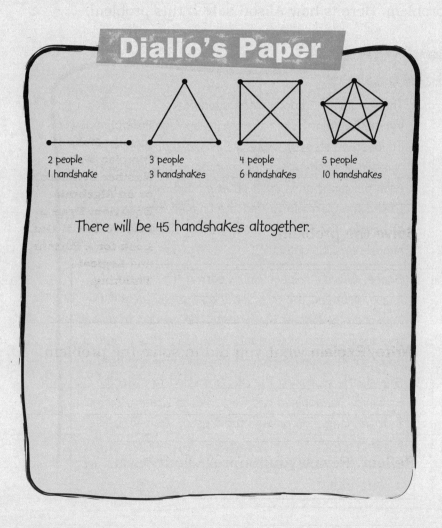

**Diallo's Paper**

2 people
1 handshake

3 people
3 handshakes

4 people
6 handshakes

5 people
10 handshakes

There will be 45 handshakes altogether.

**Score the Answer.**

According to the rubric, from **1** to **3** what score would you give Diallo? Explain why you gave that score.

_____
_____
_____
_____
_____

**Make it a 4!** Rewrite.

_____
_____
_____
_____
_____
_____
_____

Use the rubric on *page 13* to score his problem.

NOTICE: Photocopying any part of this book is prohibited by law.

# 6. Algebra

### 3. Alison Tries It.

Remember there is often more than one way to solve a problem. Here is how Alison solved this problem.

## Alison's Paper

**Keywords:** each, exactly, once, many, altogether
**Facts:** There are 10 people. Each person shakes hands with each other person exactly once.
**Question:** How many handshakes are there altogether?
**Strategy:** I Wrote an Algebraic Equation to solve the problem.
**Solve:** Let n = number of people in the room.
$(n - 1)$ = number of people each person would shake hands with
$n(n - 1)$ = number of handshakes for n people
But $n(n - 1)$ would count the handshakes between each pair of people twice. So I need to divide that formula by 2: $\frac{n(n-1)}{2}$ = number of handshakes for n people
Since there are 10 people, I will find the value of the formula when n = 10.

$$\frac{10(10-1)}{2} = \frac{10 \times 9}{2} = 45$$

There will be 45 handshakes altogether.
**Write/Explain:** I used the Write an Algebraic Equation strategy. I wrote a formula that gives the number of handshakes for a group of any size when each person shakes each other's hand exactly once. I used that formula to find the answer for a group of 10 people. The formula showed there would be 45 handshakes altogether.

**Score:** Alison's solution would earn a **4** on a test. Alison identified the question that was asked and the keywords. She knew which facts she needed to answer the question and which facts she didn't. She picked a good strategy and explained why and how she used it. She clearly explained the steps needed to solve the problem. She labeled her work. Alison wrote a formula that would give an answer for a group of any size, which she used to find the answer for a group of 10 people. Alison did everything that was required to get a score of **4**.

# 6. Algebra

### 4. Answers to Parts 1 and 2.

## Guided Problem #1

The 10 members of the Debating Club are in a room. Five are wearing green shirts. Five are wearing blue shirts. Each member shakes hands with each other exactly once. How many handshakes will there be altogether?

Keywords: ??

**1. Try It Yourself. (page 52)**

**Question:** How many handshakes will there be altogether?

**Keywords:** each, exactly, once, many, altogether

**Facts:** There are 10 club members. Each of them shakes hands with each other exactly once.

**Strategy:** Make It Simpler, Look For a Pattern

**Solve:** I can Solve a Simpler Problem. Then I can Look For a Pattern from that simpler problem to find the solution to the original problem.

| People Shaking Hands | Number of People | Combinations | Number of Handshakes |
|---|---|---|---|
| A, B | 2 | AB, ~~BA~~ | 1 |
| A, B, C | 3 | AB, AC, ~~BA~~, BC, ~~CA~~, ~~CB~~ | 3 |
| A, B, C, D | 4 | AB, AC, AD, ~~BA~~, BC, BD, ~~CA~~, ~~CB~~, CD, ~~DA~~, ~~DB~~, ~~DC~~ | 6 |
| A, B, C, D, E | 5 | AB, AC, AD, AE, ~~BA~~, BC, BD, BE, ~~CA~~, ~~CB~~, CD, CE, ~~DA~~, ~~DB~~, ~~DC~~, DE, ~~EA~~, ~~EB~~, ~~EC~~, ~~ED~~ | 10 |

↓ +2
↓ +3
↓ +4

The table shows that the change in the number of handshakes increases by 1 more each time you add a person to the group. I can extend the pattern until I have the total number of handshakes for 10 people.

NOTICE: Photocopying any part of this book is prohibited by law.

# 6. Algebra

| Number of People | 2 | 3 | 4 | 5 | 6 | 7 | 8 | 9 | 10 |
|---|---|---|---|---|---|---|---|---|---|
| Number of Handshakes | 1 | 3 | 6 | 10 | 15 | 21 | 28 | 36 | 45 |

→ +2  → +3  → +4  → +5  → +6  → +7  → +8  → +9

If 10 club members shake hands with each of the other club members exactly once, there will be a total of 45 handshakes.

**Write/Explain:** First I gathered my facts and realized that the information about the shirt colors was not needed to solve this problem. Then I **Solved a Simpler Problem** and used my results to answer the question. First, I found the total number of handshakes for groups of 2, 3, 4, and 5 people. Then I **Looked For a Pattern** and extended it to 10 people. The total number of handshakes was 45.

## 2. Diallo Tries It. (page 53)

**Score the Answer:** I would give Diallo a **1**. Diallo got the correct answer, but he did not list the question that was being asked, the keywords, the facts, or the strategy. He did not write an explanation of his work.

**Make it a 4!** Rewrite.

Since Diallo provided the correct answer, you need to provide the rest of the information and an explanation of how he solved the problem.

**Question:** How many handshakes are there altogether?

**Keywords:** each, exactly, once, altogether

**Facts:** There are 10 people. Each person shakes hand with each other person exactly once.

**Strategies:** Draw Pictures, Look for a Pattern.

**Solve:** I can extend the pattern I found in the picture.
$1 \to 3 \to 6 \to 10 \to 15 \to 21 \to 28 \to 36 \to 45$

+2  +3  +4  +5  +6  +7  +8  +9

**Write/Explain:** I **Drew Pictures** to find the numbers of handshakes for groups of 2, 3, 4, and 5 people. Then I **Looked For a Pattern** in those answers. The pattern shows that each time you increase the number of people in the group by 1, the change in the number of handshakes increases by 1. I used the pattern to find how many handshakes there would be in a group of 10 people.

# 6. Algebra

## Guided Problem #2

At Lake Logan, there are two boat-rental companies. Pier Rentals charges a fee of $20.00 plus an hourly rate of $5.00 to rent a sailboat. Lakefront Boats charges a fee of $12.50 plus an hourly rate of $7.50. How long would a boat rental have to be for it to be cheaper to rent from Pier Rentals than from Lakefront Boats?

Keywords: ? ?

### 1. Try It Yourself.

Answer the questions below to get a score of **4**.

What **question** are you being asked?

What are the **keywords**?

What are the **facts** you need to solve the problem?

What **strategy** can you use to solve the problem?

**Hint:** Possible answers include **Make a Table** and **Draw a Picture.**

**Solve** the problem.

**Write/Explain** what you did to solve the problem.

**Reflect.** Review and improve your work.

# 6. Algebra

## 2. Margie Tries It.

### Margie's Paper

**Question:** How long would a boat rental have to be for it to be cheaper to rent from Pier Rentals than from Lakefront Boats?

**Keywords:** fee, hourly rate, long, cheaper

**Facts:** Pier Rentals charges a fee of $20 plus an hourly rate of $5. Lakefront Boats charges a fee of $12.50 plus an hourly rate of $7.50.

**Strategy:** I Make a Table.

**Solve:**

| Number of Hours Rented | 0 | 1 | 2 | 3 | 4 | 5 |
|---|---|---|---|---|---|---|
| Total Charge for Pier Rentals | $20.00 | $25.00 | $30.00 | $35.00 | $40.00 | $45.00 |
| Total Charge for Lakefront Boats | $12.50 | $20.00 | $27.50 | $35.00 | $42.50 | $50.00 |

When you rent for more than 4 hours, Pier Rentals will be cheaper than Lakefront Boats.

**Write/Explain:** I Made a Table to show how much it would cost to rent from each place. I started with the fee for each company and added the hourly charge for each hour. The table shows that Pier Rentals starts being cheaper than Lakefront Boats after 4 hours.

### Score the Answer.

According to the rubric, from **1** to **3** what score would you give Margie? Explain why you gave that score.

_____
_____
_____
_____
_____
_____
_____

**Make it a 4!** Rewrite.

_____
_____
_____
_____
_____
_____
_____
_____
_____

*Use the rubric on page 13 to score her problem.*

# 6. Algebra

### 3. Bruce Tries It.

Remember there is often more than one way to solve a problem. Here is how Bruce solved this problem.

## Bruce's Paper

**Question:** How long would a boat rental have to be for it to be cheaper to rent from Pier Rentals than from Lakefront Boats?

**Keywords:** fee, hourly rate, long, cheaper

**Facts:** Pier Rentals charges a fee of $20 plus an hourly rate of $5. Lakefront Boats charges a flat fee of $12.50 plus an hourly rate of $7.50.

**Strategy:** Write Equations

**Solve:** I can Write Equations to represent the amounts charged by each boat rental company. Then I Make a Table that compares the charges.

total charge = fee + (hourly rate)(number of hours)

Let $x$ = number of hours, and $y$ = total charge.

For Pier Rentals, $y = 20 + 5x$.

For Lakefront Boats, $y = 12.5 + 7.5x$

I can Make a Table for each equation to compare the amounts charged by each company.

Pier Rentals will be cheaper than Lakefront Boats for all rentals that are longer than 3 hours.

| Number of Hours (x) | 0 | 1 | 2 | 3 | 4 | 5 |
|---|---|---|---|---|---|---|
| Pier Rentals (y) | 20 | 25 | 30 | 35 | 40 | 45 |
| Lakefront Boats (y) | 12.5 | 20 | 27.5 | 35 | 42.5 | 50 |

**Write/Explain:** I Wrote Algebraic Equations to represent the amounts charged by each boat rental company. Then I Made a Table to compare the charges. Pier Rentals will be cheaper than Lakefront Boats for all rentals that are longer than 3 hours.

**Score:** Bruce's solution would earn a 4 on our rubric. He identified the question that was asked, the keywords, the facts, and picked good strategies. He explained why he chose them, and how he used them. Then he clearly explained the steps he took to solve the problem. He also labeled his work.

NOTICE: Photocopying any part of this book is prohibited by law.

# 6. Algebra

### 4. Answers to Parts 1 and 2.

## Guided Problem #2

At Lake Logan, there are two boat-rental companies. Pier Rentals charges a fee of $20.00 plus an hourly rate of $5.00 to rent a sailboat. Lakefront Boats charges a fee of $12.50 plus an hourly rate of $7.50. How long would a boat rental have to be for it to be cheaper to rent from Pier Rentals than from Lakefront Boats?

**Keywords:** ??

### 1. Try It Yourself. (page 57)

**Question:** How long would a boat rental have to be for it to be cheaper to rent from Pier Rentals than from Lakefront Boats?

**Keywords:** fee, hourly rate, cheaper

**Facts:** Pier Rentals charges a fee of $20 plus an hourly rate of $5. Lakefront Boats charges a fee of $12.50 plus an hourly rate of $7.50.

**Strategies:** Write Algebraic Equations; Draw a Graph

**Solve:** I can Write Algebraic Equations to represent the amounts charged by each boat rental company. Then I can Draw a Graph that represents each of the equations.

total charge = flat fee +(hourly rate)(number of hours)
Let $x$ = number of hours, and $y$ = total charge.

For Pier Rentals, $y = 20 + 5x$
For Lakefront Boats, $y = 12.5 + 7.5x$

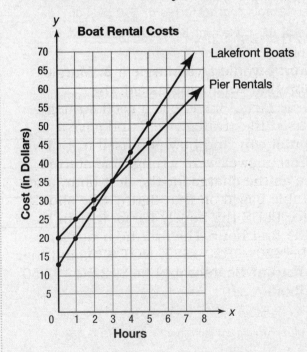

Pier Rentals will be cheaper than Lakefront Boats for all rentals of more than 3 hours.

# 6. Algebra

**Write/Explain:** I **Wrote Algebraic Equations** to represent the amounts charged by each boat rental company. Then I used the **Draw a Graph** strategy to compare how much it would cost to rent boats from Pier Rentals and Lakefront Rentals for different numbers of hours. The graph shows that for rentals of less than 3 hours, the charge for Lake Front Boats will be less than Pier Rentals. At 3 hours, the cost will be the same for both companies. For rentals of more than 3 hours, the charge for Pier Rentals will be less than Lakefront Boats.

## 2. Margie Tries It. (page 58)

**Score the Answer:** I would give Margie a **3**. Margie gave the question that was being asked, the keywords, and the facts. She chose a good strategy, **Make a Table,** used the strategy well, and gave a good explanation of why and how she used it. But Margie's table and answer were wrong. She started with just the fee as the charge for the first hour, and completed the table based on that figure. She should have used the fee PLUS the hourly fee for 1 hour to get the fee for the first hour. The first hour at Pier Rentals would be $20 + $5 = $25, NOT $20, and the first hour at Lakefront Boats would be $12.50 + 7.50 = $20, NOT $12.50.

**Make it a 4!** Rewrite.

Correct Margie's table and use it to correct her answer.

| Number of Hours Rented | 0 | 1 | 2 | 3 | 4 | 5 |
|---|---|---|---|---|---|---|
| Total Charge for Pier Rentals | $20.00 | $25.00 | $30.00 | $35.00 | $40.00 | $45.00 |
| Total Charge for Lakefront Boats | $12.50 | $20.00 | $27.50 | $35.00 | $42.50 | $50.00 |

When you rent for more than 3 hours, Pier Rentals will be cheaper than Lakefront Boats.

NOTICE: Photocopying any part of this book is prohibited by law.

# 6. Algebra

## Guided Problem #3

The graph shows the cost of Internet access at The WiFi Internet Cafe.

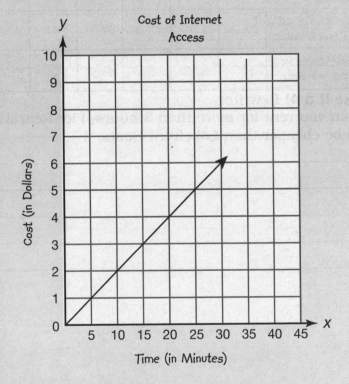

Based on the graph, how much would it cost to surf the Internet for 45 minutes?

Keywords: ? ?

### 1. Try It Yourself.

Answer the questions below to get a score of **4**.

What **question** are you being asked?

_____
_____
_____

What are the **keywords**?

_____

What are the **facts** you need to solve the problem?

_____
_____

What **strategy** can you use to solve the problem?

_____
_____

> **Hint**
> Possible answers include **Make a Table, Logical Thinking, Divide and Conquer,** and **Draw a Picture**.

**Solve** the problem.

_____
_____
_____

**Write/Explain** what you did to solve the problem.

_____
_____

**Reflect.** Review and improve your work.

_____

NOTICE: Photocopying any part of this book is prohibited by law.

# 6. Algebra

## 2. Felipe Tries It.

### Felipe's Paper

**Question:** Find the slope for two points on the line: (20, 4), (30, 6).

**Keywords:** graph, cost, much

**Facts:** The graph shows how much The WiFi Internet Cafe charges for Internet access for up to 30 minutes.

**Strategy:** Logical Thinking

**Solve:** The slope of the line gives the rate per minute. Find the slope for 2 points in the line (20,4) (30,6).

$$\text{Slope} = \frac{\text{change in y}}{\text{change in x}}$$

$$= \frac{6 - 4}{30 - 20}$$

$$= \frac{2}{10} = 0.2$$

The WiFi Internet Cafe charges a rate of $0.20 per minute for Internet access.

**Write/Explain:** I used Logical Thinking to find the answer. The graph is a line in which the total cost rises as the number of minutes rises. That means the slope of the line is the cost per minute. So, I chose two points on the line and used them to find the slope. The slope is 0.2 dollar, or $0.20. The cost per minute is $0.20.

**Score the Answer.**

According to the rubric, from **1** to **3** what score would you give Felipe? Explain why you gave that score.

_____
_____
_____
_____
_____

**Make it a 4!** Rewrite.

_____
_____
_____
_____
_____
_____
_____
_____
_____

**Use the rubric on *page 13* to score his problem.**

NOTICE: Photocopying any part of this book is prohibited by law.

# 6. Algebra

### 3. Celeste Tries It.

Remember, there is often more than one way to solve a problem. Here is how Celeste solved this problem.

## Celeste's Paper

**Question:** How much would it cost to surf the Internet for 45 minutes?
**Keywords:** graph, cost, much
**Facts:** The graph shows the cost of Internet access at The WiFi Internet Cafe.
**Strategies:** Make a Table and Look for a Pattern
**Solve:** I can use the values in the graph to Make a Table. If there is a Pattern, I can use it to find the answer.

| Minutes | 5 | 10 | 15 | 20 | 25 | 30 | 35 | 45 |
|---|---|---|---|---|---|---|---|---|
| Total Charge | $1 | $2 | $3 | $4 | $5 | $6 | $7 | $9 |

It will cost $9.00 to surf the Internet for 45 minutes.
**Write/Explain:** I Made a Table and Looked For a Pattern. I used the graph to find the costs for 5, 10, 15, 20, 25, and 30 minutes. I listed costs and times in a table. I saw that the cost increased $1 every 5 minutes. I extended that pattern to find the cost for 45 minutes. According to that pattern, 45 minutes would cost $9.00.

**Score:** We would give Celeste a score of 4. She listed the question, Keywords, and facts. She chose good strategies, explained how she used them, and showed that she understood the problem. Her solution process was clear and she answered the question. She labeled her work. Celeste's work was perfect!

# 6. Algebra

### 4. Answers to Parts 1 and 2.

## Guided Problem #3

The graph shows the cost of Internet access at The WiFi Internet Cafe.

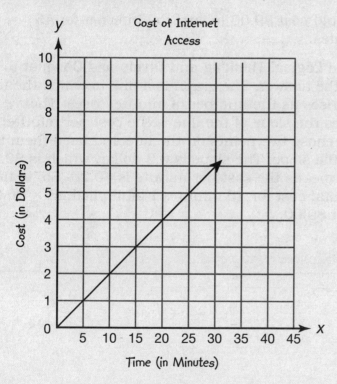

Based on the graph, how much would it cost to surf the Internet for 45 minutes?

Keywords:

### 1. Try It Yourself. (page 62)

**Question:** How much would it cost to surf the Internet for 45 minutes?

**Keywords:** graph, cost

**Facts:** The graph shows the cost of Internet access at The WiFi Internet Cafe.

**Strategy:** Divide and Conquer

**Solve:** First, find the slope of the line. That will give you the cost per minute.

Choose two points: (10, 2) and (30, 6).

$$\text{Slope} = \frac{\text{rise}}{\text{run}}$$
$$= \frac{6 - 2}{30 - 10}$$
$$= \frac{4}{20}$$
$$= 0.2$$

So, the cost per minute is $0.20.

Cost for 45 minutes = 45 × $0.20 = $9.00

**Write/Explain:** I used the **Divide and Conquer** strategy. *First,* I found the slope of the line. The slope of that line is the cost of Internet access per minute. *Second,* when I found that slope, I multiplied the cost per minute by 45 to find the cost of 45 minutes of Internet access. The cost is $9.00.

NOTICE: Photocopying any part of this book is prohibited by law.

# 6. Algebra

## 2. Felipe Tries It. (page 63)

**Score the Answer:** I would give Felipe a **2**. Felipe listed the keywords and the facts. He chose a good strategy, used the strategy well, and gave a good explanation of his work. But Felipe did not list the question asked. As a result, he answered the wrong question: Felipe gave the rate per minute instead of the cost for 45 minutes.

**Make it a 4!** Rewrite.

**Question:** How much would it cost to surf the Internet for 45 minutes?

Internet access = $0.20 per minute

Internet access for 45 minutes = 45 × $0.20 = $9.00

It would cost $9.00 to surf the Internet for 45 minutes.

I used **Logical Thinking** and **Divide and Conquer** to find the answer. The graph is a line in which the total cost rises as the number of minutes rises. That means the slope of the line is the cost per minute. So, I chose two points on the line and used them to find the slope. The slope is 0.2 dollar, which is $0.20. That means the cost per minute is $0.20. So, to find the total cost for 45 minutes I multiplied 45 × $0.20 to get $9.00.

# 6. Algebra

# Quiz Problems

Here are some problems for you to try. Keep your **rubric** handy while you solve the problem. Let's see if you can score a **4**.

**1.** Paul worked at the Treetorn Paper Company for 3 weeks. Each week he worked 1 more hour than the previous week. During these 3 weeks he worked a total of 96 hours. How many hours did he work each week?

**2.** A bus begins a 420-kilometer trip at 2:30 p.m. The bus is scheduled to reach its destination at 8:30 p.m.. During the first 2 hours, the bus is in heavy traffic and has an average speed of only 30 kilometers per hour. What average speed must the bus travel during the next 4 hours in order to reach its destination on time?

**3.** An 80-gallon water tank begins to leak at a constant rate. The graph below shows the amount of water that is left in the tank.

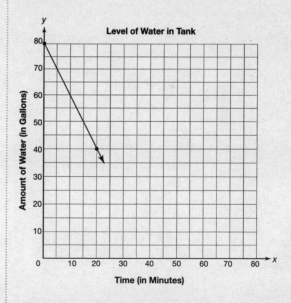

If the tank continues to leak at the same rate, how many minutes will it take for the tank to empty?

# 6. Algebra

**4.** The table lists earnings information for one month for some sales representatives at Nextco Products.

| A | B | C | D | E |
|---|---|---|---|---|
| Sales Representatives | Base Salary | Sales | Commission | Total Earnings |
| Alvarez, Aldo | $3,200 | $61,000 | $3,660 | $6,860 |
| Thorhill, Roger | $2,800 | $44,000 | $2,640 | $5,440 |
| Vermon, Aida | $3,500 | $70,000 | $4,200 | $7,700 |

How is the commission related to the sales? Write an equation to explain your answer.

**5.** A souvenir stand sells large flags for $9.00 each and small flags for $6.00 each. The souvenir stand sold 10 flags for a total of $72.00. How many of each kind of flag did the souvenir stand sell?

**6.** The figures below show the first four numbers in a sequence of rectangular numbers.

```
              . . .    . . . .    . . . . .
   . .   . . .    . . . .    . . . . .
   . .   . . .    . . . .    . . . . .
         . . .    . . . .    . . . . .
    2     6        12          20
```

What would be the sixth rectangular number in this sequence?

**7.** Jed has 18 coins. Some are quarters and the rest are dimes. The total value of the coins is $3.30. How many quarters and how many dimes does he have?

**Mathematics Open-Ended Questions, Level G**

# 7. Geometry

When you study geometry, you study **points**, **lines**, and **shapes**. You study where they are and how they relate to each other. When you describe two streets as parallel streets, you are using geometry. When you make an enlargement of a photograph, you are creating similar geometric figures. If you figure out how much wood it might take to make a box, that's also part of geometry.

# 7. Geometry

Here is a geometry problem that might be on your tests. Let's look at one way of solving the problem that would get a perfect score of **4** according to our **rubric**.

## Modeled Problem

Jason has a **rectangular** photograph that is 4.5 inches by 6 inches. He makes an enlargement in which one of the **dimensions** is 27 inches. What could be the dimensions of the **enlargement**? Find all of the possibilities.

**Keywords: rectangular, dimensions, enlargement**

### 1. Read and Think

Let's **read** the problem.

What **question** are we asked?

- We are asked to find all of the possibilities for the dimensions of the enlargement.

Do we recognize what the **keywords** are?

- Yes. rectangular, dimensions, enlargement

What **facts** are we given?

- The dimensions of a rectangular photograph are 4.5 inches by 6 inches.
- In an enlargement of the original photograph, one of the dimensions is 27 inches.

### 2. Select a Strategy

The problem describes a situation in which there is a pair of similar figures: the original figure and the enlargement. We can **Draw a Picture** to show how the sides of these two figures could correspond.

### 3. Solve

The 27-inch side of the enlargement could correspond to the 4.5-inch side of the original figure.

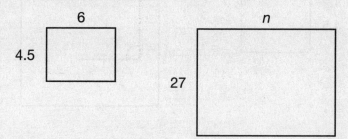

Since the original and the enlargement are similar, their corresponding sides are proportional. So, we can write and solve a proportion to find the missing dimension.

# 7. Geometry

$$\frac{\text{original} \to 4.5}{\text{enlargement} \to 27} = \frac{6 \leftarrow \text{original}}{n \leftarrow \text{enlargement}}$$

$$\frac{4.5}{27} = \frac{6}{n}$$

$$4.5n = 27(6)$$

$$4.5n = 162$$

$$\frac{4.5n}{4.5} = \frac{162}{4.5}$$

$$n = 36$$

The missing side of the enlargement could be 36 inches. So, the dimensions of the enlargement could be 27 inches by 36 inches.

Another possibility is that the 27-inch side of the enlargement could correspond to the 6-inch side of the original. Again, we can write and solve a proportion to find the missing dimension.

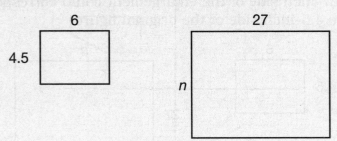

$$\frac{6}{27} = \frac{4.5}{n}$$

$$6n = 27(4.5)$$

$$6n = 121.5$$

$$\frac{6n}{6} = \frac{121.5}{6}$$

$$n = 20.25$$

The missing side could be 20.25 inches. The dimensions of the enlargement could be 27 inches by 20.25 inches.

So, there are two possibilities for the dimensions of the enlargement: 27 inches by 36 inches, and 20.25 inches by 27 inches.

### 4. Write/Explain

To solve this problem, we started with both dimensions of one rectangle and one dimension of a similar rectangle that was an enlargement. We Used Drawings to show two different enlargements that were possible. Then we used proportions to find the missing side in each of these enlargements.

### 5. Reflect

Let's review our work and answer.

- Did we show that we knew what the problem asked?  **Yes. We found two possible sets of dimensions for the enlargement.**
- Did we know what the keywords were?  **Yes.**
- Did we show that we knew what facts were given?  **Yes.**
- Did we name and use the correct strategy?  **Yes.**
- Was our math correct?  **Yes. We checked it. It was correct.**
- Did we label our work?  **Yes.**

# 7. Geometry

- Was our answer correct? **Yes.**
- Did we explain why we chose the strategy and how it was used to get the right answer? **Yes. It helped us visualize the possible ways in which the enlargement could correspond to the original rectangle.**
- Were all of our steps included? **Yes.**
- Did we write a good, clear explanation of our work? **Yes. We explained each step in the solution process.**

This solution would earn a **4** on our rubric. The answer is correct and the solution process is clearly explained.

*Use the rubric on page 13 to score this work.*

On the following pages are some **Guided Open-Ended Math Problems.** For each problem there are **four parts**. In the **first part**, you will solve the problem with guided help. In the **second part**, you will score and correct a solution with guided help. The **third part** shows one solution that scores a perfect **4**. This solution may or may not differ from your way. The **fourth part** has *answers* to the **first** and **second parts** so you can check your work.

## Guided Problem #1

What two transformations could you use to move figure *F* onto figure *G* without moving the figure through Quadrant II? Describe each transformation as specifically as you can.

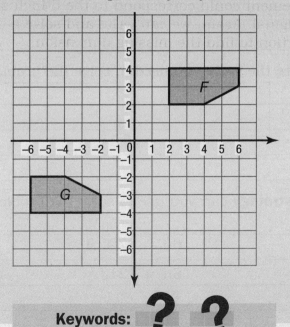

Keywords: ? ?

NOTICE: Photocopying any part of this book is prohibited by law.

# 7. Geometry

## 1. Try It Yourself.

Answer the questions below to get a score of **4**.

What **question** are you being asked?

_____
_____
_____
_____

What are the **keywords**?

_____
_____
_____
_____

What are the **facts** you need to solve the problem?

_____
_____
_____
_____

What **strategy** can you use to solve the problem?

_____
_____
_____

**Solve** the problem.

_____
_____
_____
_____
_____

> **Hint**
> Possible answers include: **Draw a Picture, Work Backward, Logical Thinking**, and **Guess and Test**.

**Write/Explain** what you did to solve the problem.

_____
_____
_____
_____
_____

**Reflect.** Review and improve your work.

_____
_____
_____
_____
_____
_____

# 7. Geometry

## 2. Emily Tries It.

### Emily's Paper

**Question:** What two transformations could move figure F onto figure G?

**Facts:** The vertices of Figure F are (2, 2), (2, 4), (6, 4), (6, 3), and (4, 2). The vertices of figure G are (−6, −2), (−6, −4), (−2, −4), (−2, −3), (−4, −2).

**Strategy:** Draw a Picture

**Solve:** I can Draw a Picture to help me solve the problem.

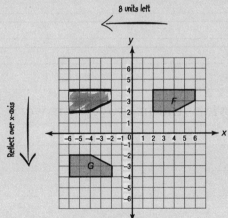

**Write/Explain:** I saw that if I translated figure F 8 units to the left, I could reflect it over the x-axis to put it on figure G. I used a picture to make sure that I was right. First, I drew the translation. Then I made sure that each point in the translation and the corresponding point in figure G were the same distance from the x-axis. That meant that figure G was a reflection of the figure I drew.

Use the rubric on *page 13* to score her problem.

# 7. Geometry

**Score the Answer.**

According to the rubric, from **1** to **3** what score would you give Emily? Explain why you gave that score.

_____
_____
_____
_____
_____
_____

**Make it a 4!** Rewrite.

_____
_____
_____
_____
_____
_____
_____
_____
_____

NOTICE: Photocopying any part of this book is prohibited by law.

# 7. Geometry

## 3. Bernard Tries It.

Remember there is often more than one way to solve a problem. Here is how Bernard solved this problem.

### Bernard's Paper

**Question:** What two transformations could you use to move figure F onto figure G without moving the figure through Quadrant II?
**Keywords:** transformation, specifically, Quadrant II
**Facts:** I know the vertices of figure F and figure G.
**Strategies:** Draw a Picture and Guess and Test.
**Solve:** First, I can try translating figure F to Quadrant IV so that the bottom of the translated figure is aligned with the bottom of figure G.

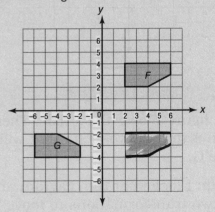

I can't see a way to move the shape I drew so that it fits onto figure G. I will try another solution. First, I will reflect figure F over the x-axis.

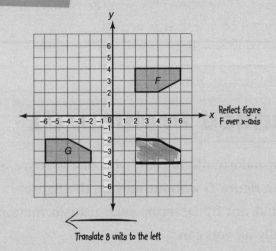

Reflect figure F over x-axis

Translate 8 units to the left

If you translate the reflection that I drew in Quadrant IV 8 units to the left, it moves onto figure G. So, to move figure F onto figure G, you reflect it over the x-axis. Then you translate it 8 units to the left.

**Write/Explain:** I used the Draw a Picture and the Guess and Test strategies. First, I tried translating figure F into Quadrant IV. But I couldn't see a way to move the shape I drew onto figure G. So, I tried reflecting Figure F over the x-axis. When I did that, I saw that I could translate my drawing 8 units to the left to put it on figure G. (cont.)

NOTICE: Photocopying any part of this book is prohibited by law.

# 7. Geometry

(Bernard's paper cont.)

**Score:** Bernard identified the question that was asked, the keywords, and facts. Bernard chose good strategies and gave a clear explanation of how he used them to find the solution. He labeled his work. Bernard's solution would earn a **4** on a test. It is perfect!

### 4. Answers to Parts 1 and 2.

## Guided Problem #1

What two transformations could you use to move figure F onto figure G without moving the figure through Quadrant II? Describe each transformation as specifically as you can.

**Keywords:** ? ?

### 1. Try It Yourself. (pages 73–74)

**Question:** What two transformations could you use to move figure F onto figure G without moving the figure through Quadrant II?

  **Keywords:** transformation, Quadrant II, specifically

  **Facts:** The vertices of figure F are (2, 2), (2, 4), (6, 4), (6, 3), and (4, 2). The vertices of figure G are (−6, −2), (−6, −4), (−2, −4), (−2, −3), and (−4, −2).

**Strategy:** Draw a Picture

**Solve:**

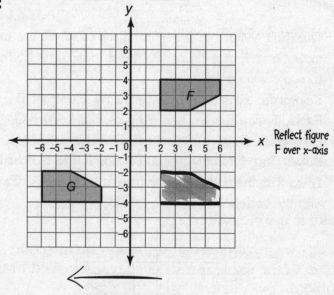

First, reflect figure F over the x-axis. Then translate the reflection 8 units to the left. That will move figure F onto figure G.

**Write/Explain:** I looked at the diagram and visualized how to use two transformations to move figure F onto figure G. Then I **Drew a Picture** to show how to move the figure. First, I reflected figure F over the x-axis. I made sure that each point in the original figure and the corresponding point in the reflection were the same distance from the x-axis. Then I counted the

# 7. Geometry

number of units each point in the translation needed to move left in order move the corresponding point in figure G.

### 2. Emily Tries It. (page 75)

**Score the Answer:** I would give Emily a **2**.
Emily listed the facts. She chose a good strategy, explained why and how she used it, and gave a clear explanation of her work. But she forgot the keywords and one part of the question. The question asks you to move figure F onto figure G without moving through Quadrant II. Emily forgot that the figure could not move through Quadrant II. As a result, she gave the wrong answer.

**Make it a 4!** Rewrite.

First, list the keywords and write the question so that it is the same question that was asked in the problem.

**Question:** What two transformations could you use to move figure F onto figure G without moving the figure through Quadrant II?

**Keywords:** transformation, Quadrant II, specifically
The next thing that needs to be changed is the solution. Figure F must move onto figure G without moving through Quadrant II. **Draw a Picture** to show how to do this, and describe the two transformations in the drawing.

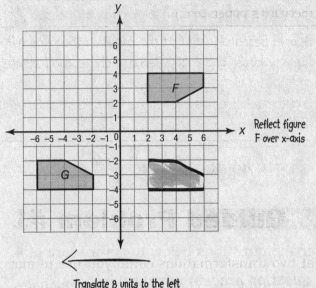

First, reflect figure F over the x-axis. Then translate the reflection 8 units to the left. That will move figure F onto figure G.

Next, change Emily's explanation so that it matches the new solution.

I looked at the diagram and visualized how to move figure F onto figure G. I **Drew a Picture** to show how to move the figure. First, I reflected figure F over the x-axis. I made sure that each point in the original figure and the corresponding point in the reflection were the same distance from the x-axis. Then I counted the number of units each point in the translation needed to move left in order to move the corresponding point in figure G.

NOTICE: Photocopying any part of this book is prohibited by law.

# 7. Geometry

## Guided Problem #2

A rectangle with a perimeter of 36 centimeters was cut into two squares. What were the dimensions of the original rectangle?

Keywords: ? ?

### 1. Try It Yourself.

Answer the questions below to get a score of **4**.

What **question** are you being asked?

_____
_____
_____

What are the **keywords**?

_____
_____
_____

What are the **facts** you need to solve the problem?

_____
_____
_____

What **strategy** can you use to solve the problem?

_____
_____
_____

> **Hint**
> Possible answers include: **Work Backward, Draw a Picture, Logical Thinking**, and **Make an Organized List** or **Table**.

**Solve** the problem.

_____
_____
_____
_____

**Write/Explain** what you did to solve the problem.

_____
_____
_____
_____

**Reflect.** Review and improve your work.

_____
_____
_____

NOTICE: Photocopying any part of this book is prohibited by law.

# 7. Geometry

### 2. Chun Tries It.

**Chun's Paper**

**Keywords:** rectangle, squares, perimeter, dimensions

**Facts:** The rectangle is cut into two squares.
The perimeter of a square is 36 centimeters.

**Solve:**

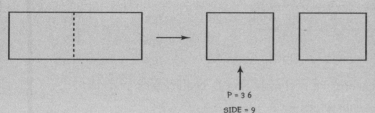

P = 36
SIDE = 9

P = 36 cm
side = 9 cm

So, each square has sides of 9 cm. If you put them back together, you get a rectangle that is 18 cm long and 9 cm wide.

**Write/Explain:** I broke the rectangle into two squares. I found the side length of each square. That told me the dimensions of the rectangle.

**Score the Answer.**

According to the rubric, from **1** to **3** what score would you give Chun? Explain why you gave that score.

_____
_____
_____
_____
_____
_____

**Make it a 4!** Rewrite.

_____
_____
_____
_____
_____
_____
_____

Use the rubric on *page 13* to score his problem.

NOTICE: Photocopying any part of this book is prohibited by law.

# 7. Geometry

## 3. Natalie Tries It.

Remember there is often more than one way to solve a problem. Here is how Natalie solved this problem.

### Natalie's Paper

**Question:** What were the dimensions of the original rectangle?

**Keywords:** rectangle, squares, perimeter, dimensions

**Facts:** A rectangle with a perimeter of 36 centimeters was cut into two squares.

**Strategies:** Draw a Picture and Make an Organized List

**Solve:**
The picture shows a rectangle that is made of two congruent squares. Each side of each square has the length, $s$.

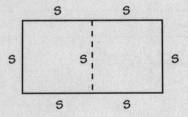

Length of the rectangle = $2s$
Width of the rectangle = $s$
The picture shows that the rectangle is twice as long as it is wide. So, I can look for a rectangle in which the length is twice the width, and which has a perimeter of 36 centimeters. I can make an organized list.

| Width (W) | Length (L) | Perimeter (P) $2(L + W) = P$ | Is the length twice the width? | Is the perimeter 36 inches? |
|---|---|---|---|---|
| 4 | 8 | $2(8 + 4) = 24$ | Yes | No |
| 5 | 10 | $2(10 + 5) = 30$ | Yes | No |
| 6 | 12 | $2(12 + 6) = 36$ | Yes | Yes |

The original rectangle was 12 centimeters long and 6 centimeters wide.

**Write/Explain:** I Drew a Picture to represent a rectangle that is made of two squares. The picture showed that one side of that rectangle must be twice as long as the other side. I also knew that the perimeter of the rectangle was 36 centimeters. So, I Made an Organized List to find a rectangle that had a perimeter of 36 inches and that had a length that was twice as long as the width.

**Score:** Natalie's solution would earn a **4** on a test. Natalie identified question that was asked, the keywords, and the facts. Natalie chose good strategies, explained why and how she used them, and gave a clear explanation of the steps she took in order to solve the problem. She also labeled her work.

# 7. Geometry

### 4. Answers to Parts 1 and 2.

## Guided Problem #2

A rectangle with a perimeter of 36 centimeters was cut into two squares. What were the dimensions of the original rectangle?

Keywords:

**1. Try It Yourself. (page 80)**

**Question:** What were the dimensions of the original rectangle?

**Keywords:** rectangle, squares, perimeter, dimensions

**Facts:** A rectangle with a perimeter of 36 centimeters was cut into two squares.

**Strategies:** Work Backward and Draw a Picture

**Solve:**

Start with two squares. Each has sides with length $x$.

Put the squares together to make a rectangle. The picture shows that the perimeter of the rectangle is equal to $6x$.

$$36 \text{ cm} = 6x$$
$$\frac{36 \text{ cm}}{6} = \frac{6x}{6}$$
$$6 \text{ cm} = x$$

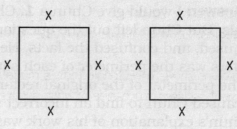

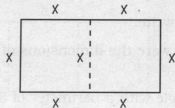

Use the Drawing and the value of $x$ to find the dimensions of the rectangle.

width of rectangle = $x$ = 6 cm

length of rectangle = $2x$ = 2(6 cm) = 12 cm

The dimensions of the original rectangle are 12 centimeters by 6 centimeters.

# 7. Geometry

**Write/Explain:** I **Worked Backward** and **Drew a Picture**. I drew two congruent squares that had sides of x. I combined the squares to make a rectangle with a perimeter of 6x. Then I used the perimeter to Write an Algebraic Equation that would give the value of x. Then I found the length and width of the original rectangle. As the diagram shows, x, which is equal to 6 cm, is the width of the rectangle. So the length, 2x, is 12 cm.

## 2. Chun Tries It. (page 81)

**Score the Answer:** I would give Chun a **1**. Chun gave the keywords. But Chun left out the question, the strategy he used, and confused the facts. He thought 36 centimeters was the perimeter of each square, instead of the perimeter of the original rectangle. This confusion caused Chun to find an incorrect solution. In addition, Chun's explanation of his work was unclear.

**Make it a 4!** Rewrite.

**Question:** What were the dimensions of the original rectangle?

**Facts:** A rectangle with a perimeter of 36 centimeters was cut into two squares.

Start with two squares.
  Each has a side with length x.

  Put the squares together to make a rectangle.
    The picture shows that perimeter of the rectangle is equal to 6x.

$$36 \text{ cm} = 6x$$

$$\frac{36 \text{ cm}}{6} = \frac{6x}{6}$$

$$6 \text{ cm} = x$$

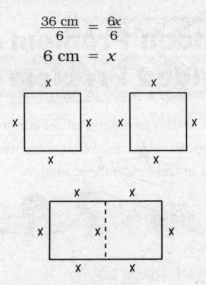

Use the diagram and the value of x to find the dimensions of the rectangle.

width of rectangle = $x$ = 6 cm

length of rectangle = $2x$ = 2(6 cm) = 12 cm

The dimensions of the original rectangle are 12 centimeters by 6 centimeters.

**Write/Explain:** I **Worked Backward** and **Drew a Picture**. I drew two congruent squares that had sides of x. I combined the squares to make a rectangle with a perimeter of 6x. Then I used the perimeter to Write an Algebraic Equation that would give the value of x. Then I found the length and width of the original rectangle. As the diagram shows, x, which is equal to 6 cm, is the width of the rectangle. So the length, 2x, is 12 cm.

# 7. Geometry

## Guided Problem #3

The diagram shows possible routes between given locations.

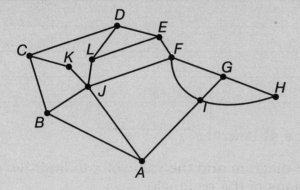

A truck driver is at point H. He must make stops at points C, E, and J. The order of the stops does not matter. What is the shortest possible route the truck driver can take?

Keywords:

### 1. Try It Yourself.

Answer the questions below to get a score of **4**.

What **question** are you being asked?
_____
_____

What are the **keywords**?
_____
_____

What are the **facts** you need to solve the problem?
_____
_____
_____

What **strategy** can you use to solve the problem?
_____
_____

**Solve** the problem.
_____
_____

**Write/Explain** what you did to solve the problem.

> **Hint**
> Possible answers include: **Logical Thinking, Make it Simpler,** and **Divide and Conquer.**

_____
_____
_____

**Reflect.** Review and improve your work.
_____
_____

# 7. Geometry

## 2. Janelle Tries It.

### Janelle's Paper

**Question:** What is the shortest possible route the driver can take?

**Keyword:** shortest

**Facts:** The driver starts at point H. He must make stops at points C, E, and J.

**Strategy:** Logical Thinking

**Solve:**
To get from H to E, the straightest path is H, G, F, E.
To get from E to C, the straightest path is E, D, C.
To get from C to J, the straightest path is C, K, J.
The shortest route the driver can take is H, G, F, E, D, C, K, and J.

**Write/Explain:** I used Logical Thinking. Since the shortest path between two points is a straight line, I looked for the straightest routes between each pair of points. That gave me a route where the driver goes through points in this order: H, G, F, E, D, C, K, and J.

**Score the Answer.**

According to the rubric, from **1** to **3** what score would you give Janelle? Explain why you gave that score.

_____
_____
_____
_____
_____
_____
_____

**Make it a 4!** Rewrite.

_____
_____
_____
_____
_____
_____
_____
_____

Use the rubric on *page 13* to score her problem.

NOTICE: Photocopying any part of this book is prohibited by law.

# 7. Geometry

### 3. Fred Tries It.

Remember there is often more than one way to solve a problem. Here is how Fred solved this problem.

## Fred's Paper

**Question:** What is the shortest possible route the driver can take?

**Keyword:** shortest

**Facts:** The driver starts at point H. He must make stops at points C, E, and J. The order in which he stops at the points does not matter.

**Strategies:** Make It Simpler and Divide and Conquer

**Solve:**

I can look at the four points on the driver's route and eliminate points that are out of the way. Since points A, D, and B are below or above the four points on the driver's route, I will try to avoid any routes that go through those points.

The driver starts at point H, which is at the right side of the diagram. The driver has to stop at point C, which is at the left side of the diagram. From right to left, the points are H, E, J, and C.

From H to E, the shortest route is H, G, F, and E.

From E to J, the shortest route is E, L, and J.

From J to C, the shortest route is J, K, and C.

The shortest route that starts at H and stops at points C, E, and J, is H, G, F, E, L, J, K, and C.

**Write/Explain:** I Made the Problem Simpler. Then I used the Divide and Conquer strategy. Firstly, I eliminated points that seemed out of the way. I found an order that went from right to left without backtracking. Then secondly, I broke the trip into parts. For each part, I looked for a route that would be as close as possible to a straight line. That gave me this route: H, G, F, E, L, J, K, and C.

**Score:** Fred identified the question that was asked, and the keywords and facts. Fred chose a good combination of strategies, explaining why and how he used them. Then he clearly explained the steps he took in order to solve the problem. He labeled his work correctly. Fred's solution was perfect. It would earn a **4** on a test.

NOTICE: Photocopying any part of this book is prohibited by law.

# 7. Geometry

## 4. Answers to Parts 1 and 2.

### Guided Problem #3

The diagram shows possible routes between given locations.

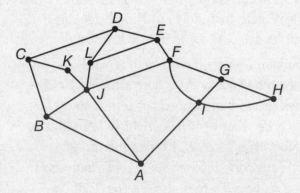

A truck driver is at point H. He must make stops at points C, E, and J. The order of the stops does not matter. What is the shortest possible route the truck driver can take?

Keywords:

### 1. Try It Yourself. (page 85)

**Question:** What is the shortest possible route the driver can take?

**Keywords:** shortest, route, points

**Facts:** A truck driver is at point H. He must make stops at points C, E, and J. The order in which he stops at the points does not matter.

**Strategy:** Logical Thinking

**Solve:** First, arrange the points so that the driver will not need to backtrack at all: H, E, J, and C.

Next, find the shortest route from point to point. Start with the route from point H to point E. The shortest route from H to E is H, G, F, and E. Now go from E to J. The only way to do this without backtracking or going further up than necessary is by going from E to L to J. Now go from J to C. The only way to do this without backtracking or going further down is by going from J to K to C.

So, the shortest route that starts at H and stops at points C, E, and J, is H, G, F, E, L, J, K, and C.

**Write/Explain:** I used **Logical Thinking**. I arranged the points in order so that the driver would not backtrack. Next, I went from point to point in the order that I listed them. That gave me this route: H, G, F, E, L, J, K, and C. Between each pair of points, I looked for a route that would be as close as possible to a straight line and would not involve going out of the way or backtracking.

# 7. Geometry

## 2. Janelle Tries It. (page 86)

**Score the Answer:** I would give Janelle a **2**. Janelle listed the question, keyword, and some of the facts. She chose a good strategy. But Janelle did not find the shortest route between the points. In addition, Janelle's explanation of why her strategy would work was unclear.

**Make it a 4!** Rewrite.

Add the missing fact to Janelle's work. Then write a solution that gives the shortest route between the points.

**Facts:** A truck driver is at point *H*. He must make stops at points *C*, *E*, and *J*. The order in which he stops at the points does not matter.

**Solve:** First, arrange the points so that the driver will not need to backtrack at all: *H*, *E*, *J*, and *C*.

Next, find the shortest route from point to point. Start with the route from point *H* to point *E*. The shortest route from *H* to *E* is *H*, *G*, *F*, and *E*. Now go from *E* to *J*. The shortest route is *E* to *L* to *J*. Now go from *J* to *C*. The shortest route is *J* to *K* to *C*.

So, the shortest route that starts at *H* and stops at points *C*, *E*, and *J*, is *H*, *G*, *F*, *E*, *L*, *J*, *K*, and *C*.

**Write/Explain: I Used Logical Thinking.** First, I arranged the points in order so that the driver would not backtrack to get from point to point. Next, I went from point to point in the order that I listed them. Between each pair of points, I looked for a route that would be as close as possible to a straight line and that would not involve going out of the way or backtracking. That gave me this route: *H*, *G*, *F*, *E*, *L*, *J*, *K*, and *C*.

# 7. Geometry

# Quiz Problems

Here are some problems for you to try. Keep your rubric handy while you solve the problem. Let's see if you can score a **4**.

**1.** Marnie is using rectangular panes of colored glass to make a decorative border for a mirror. Each pane of glass is 8 centimeters by 6 centimeters. Marnie cuts each pane along a diagonal to make two triangular pieces of glass.

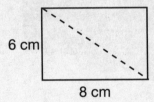

Marnie puts the diagonal edge of each pane around the edges of the mirror. The mirror is a rectangle that is 60 centimeters by 40 centimeters. How many triangular pieces of glass can fit around the mirror?

**2.** The Boxtop Group gets an order for wooden crates that have a volume of 36 cubic feet. In order to keep the cost of the crates low, the crates must use as little wood as possible. What dimensions must each crate have in order to have the least possible surface area? Assume that the dimensions of the box must be whole numbers.

**3.** How many triangles of all sizes are in the figure?

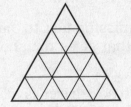

NOTICE: Photocopying any part of this book is prohibited by law.

# 7. Geometry

**4.** Elena's dog, Mike, does the same six activities in every 24-hour day: sleep, eat, chew important things, watch cartoons, bury his bone, find and dig up his bone. The circle graph, which is missing its data, shows this information. For example, every day Mike watches cartoons for about 6 hours. Which lettered sector of the graph shows the amount of time Mike spends watching cartoons each day?

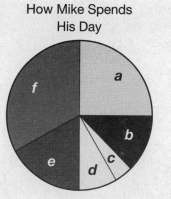

How Mike Spends His Day

**5.** What is the perimeter of the figure formed when 10 of these triangles are joined together in a row to form a parallelogram? Hint: Join the triangles as shown.

**6.** Nell has four rectangular picture frames that are 4 inches by 6 inches, 10 inches by 18 inches, 16 inches by 20 inches, and 20 inches by 30 inches. Could any two of these picture frames hold a photograph and its proportionally similar enlargement?

**7.** Twenty-seven small cubes were stacked to make a large block. Then the outside of the block was painted green. How many of the cubes have no faces painted? How many are painted on only one face? On exactly two faces? On exactly three faces?

Mathematics Open-Ended Questions, Level G

# 8. Measurement

Measurement is part of many open-ended math problems. Any problem that asks about a **size**, a **temperature**, a **period of time**, or even an **amount of money** is a measurement problem. Problems that include **rates**, such as gallons per minute, are also measurement problems.

# 8. Measurement

The tests have lots of problems that involve measurements such as perimeter, area, and volume. Let's look at a problem that involves area and see how it got a **4** on our **rubric**.

## Modeled Problem

The diagram below shows a room in which all of the corners are **right angles**. Mr. Rawlings is going to cover the floor with carpet that costs $18 **per square meter**. How **much** will it **cost** Mr. Rawlings to buy the carpet?

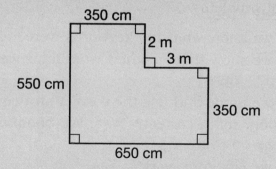

Keywords: right angles, per, square meter, much, cost

### 1. Read and Think

Let's **read** the problem.

What **question** are we asked?

- How much will it cost Mr. Rawlings to buy the carpet?

Do we recognize what the **keywords** are?

- **right angles, per, square meter, much, cost**

What **facts** are we given?

- **All of the angles in the room are right angles.**
- **The carpet costs $18 per square meter.**
- **The diagram shows the dimensions of the room.**

### 2. Select a Strategy

To solve this problem, you need to find the area of the room and the total cost of the carpet. Since the problem has several steps, we'll use the **Divide and Conquer** strategy.

### 3. Solve

Three of the dimensions are given in centimeters and two are given in meters. Since the cost of the carpet is given per square meter, write the dimensions of the room so that they are all in meters.

# 8. Measurement

$$\frac{550 \text{ cm}}{100} = 5.5 \text{ m}$$

$$\frac{350 \text{ cm}}{100} = 3.5 \text{ m}$$

$$\frac{650 \text{ cm}}{100} = 6.5 \text{ m}$$

We can think of the room as a 5.5-meter-by-6.5-meter rectangle that has 2-meter-by-3-meter rectangle missing from it.

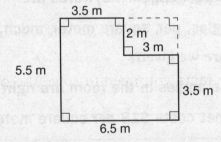

To find the area of the room, subtract the area of missing rectangle from the area of the large rectangle. Use the formula $A = lw$.

(area of large rectangle) − (area of small rectangle) = area of room

(6.5 m × 5.5 m) − (3 m × 2 m) =
35.75 m² − 6 m² = 29.75 m².

The area is 29.75 m².

Now that we know the area of the room, we can find the total cost of carpeting the room

(area in square meters) × (cost per square meter of carpet) = total cost of carpet

29.75 × $18 = $535.50

The carpet will cost $535.50.

### 4. Write/Explain

We used the Divide and Conquer and Draw a Picture strategies to solve the problem. There were 2 parts. Firstly, we converted the dimensions so they were all expressed in meters. We used the picture to help us see that the room was shaped like a large rectangle with a smaller rectangle cut out of it. To find the area, we subtracted the area of small rectangle from the area of large rectangle. Then we multiplied the number of square meters in the room by the cost per square meter of carpet. That gave us the total cost of the carpet.

### 5. Reflect

Let's review our work and answer.

- Did we show that we knew what the problem asked? **Yes.**
- Did we know what the keywords were? **Yes.**
- Did we show that we knew what facts were given? **Yes.**
- Did we name and use the correct strategy? **Yes.**
- Was our math correct? **Yes. We checked it. It was correct.**
- Did we label our work? **Yes.**
- Was our answer correct? **Yes.**
- Did we show why we used the strategy we did and how we used it? **Yes.**
- Were all of our steps included? **Yes.**
- Did we write a good, clear explanation of our work? **Yes.**

# 8. Measurement

### Score
This solution would earn a **4** on our rubric. We got the correct answer and gave a clear explanation of how we arrived at the answer.

Here are some **Guided Open-Ended Math Problems**. For each problem there are **four parts**. In the **first part**, you will solve the problem with guided help. In the **second part**, you will score and correct a solution with guided help. The **third part** shows one solution that scores a perfect **4**. This solution may or may not differ from your way. The **fourth part** has *answers* to the **first** and **second parts** so you can check your work.

## Guided Problem #1

A cardboard box that is a rectangular prism has a volume of 72 cubic feet. The base of the box is 4 feet by 3 feet. What is the surface area of the box?

Keywords:

### 1. Try It Yourself.

Answer the questions below to get a score of **4**.

What **question** are you being asked?

_____
_____

What are the **keywords**?

_____
_____
_____

What are the **facts** you need to solve the problem?

_____
_____
_____

What **strategy** can you use to solve the problem?

_____
_____
_____
_____

### Hint
Possible answers include: **Divide and Conquer, Draw a Picture,** and **Act it Out.**

Use the rubric on *page 13* to score this work.

NOTICE: Photocopying any part of this book is prohibited by law.

95

# 8. Measurement

**Solve** the problem.

_____

_____

**Write/Explain** what you did to solve the problem.

_____

_____

**Reflect.** Review and improve your work.

_____

_____

### 2. Paula Tries It.

## Paula's Paper

**Question:** What is the surface area of the box?
**Keywords:** rectangular prism, base, surface area, volume
**Facts:** The base of the box has a length of 4 feet.
The base of the box has a width of 3 feet.
**Strategy:** Divide and Conquer
**Solve:**
4 × 3 = 12  So, the area of one side of the box is 12 square feet.
6 × 12 = 72  So, the surface area of the box is 72 square feet
**Write/Explain:** I used Divide and Conquer. First I found the area of one side of the box. Since the box has 6 sides, I multiplied the area of one side by 6.

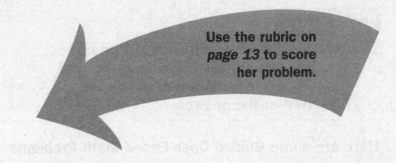

Use the rubric on *page 13* to score her problem.

**Score the Answer.**

According to the rubric, from **1** to **3** what score would you give Paula? Explain why you gave that score.

_____

_____

_____

**Make it a 4!** Rewrite.

_____

_____

_____

_____

_____

_____

NOTICE: Photocopying any part of this book is prohibited by law.

# 8. Measurement

## 3. Jerry Tries It.

Remember there is often more than one way to solve a problem. Here is how Jerry solved this problem.

### Jerry's Paper

**Question:** What is the surface area of the box?

**Keywords:** rectangular prism, volume, base, surface area

**Facts:** A box that is a rectangular prism has a volume of 72 cubic feet.
The base of the box is 4 feet by 3 feet.

**Strategies:** Act It Out, Make a Model

**Solve:** I can use cubes to make a model of the base. I will use each cube to represent a cube with sides of 1 foot each.

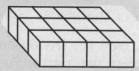

That gives me a model with 12 cubes. That's a volume of 12 cubic feet. Now I can add layers of 12 cubes until I have a figure with a volume of 72 cubes.

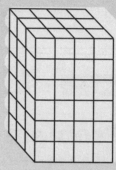

I can use the model to find the surface area of each face. There are 3 pairs of congruent faces.

4 ft × 3 ft = 12 ft²     2 × 12 ft² = 24 ft²
6 ft × 3 ft = 18 ft²     2 × 18 ft² = 36 ft²
6 ft × 4 ft = 24 ft²     2 × 24 ft² = 48 ft²
24 ft² + 36 ft² + 48 ft² = 108 ft²

The box has a surface area of 108 ft².

**Write/Explain:** I Acted It Out. I used cubes to make a base that was 4 cubes by 3 cubes. Then I added layers that were 4 cubes by 3 cubes. I did this until I had a box made of 72 cubes. I used my model to find the area of each pair of congruent faces of the cube. Then I added the areas to get the surface area.

(cont.)

# 8. Measurement

(Jerry's paper cont.)

**Score:** Jerry's solution would earn a **4** on a test. Jerry identified the question, keywords, and the facts. He picked a good strategy, and he gave a clear explanation of how he chose it and the steps he took to solve the problem. Jerry also labeled his work. Jerry's solution is perfect!

**4. Answers to Parts 1 and 2.**

## Guided Problem #1

A cardboard box that is a rectangular prism has a volume of 72 cubic feet. The base of the box is 3 feet by 4 feet. What is the surface area of the box?

**Keywords:**

**1. Try It Yourself. (pages 95–96)**

**Question:** What is the surface area of the box?

**Keywords:** rectangular prism, volume, base, surface area

**Facts:** A box that is a rectangular prism has a volume of 72 cubic feet.

The base of the box has a length of 4 feet.

The base of the box has a width of 3 feet.

**Strategy:** Divide and Conquer

**Solve:**

Volume of a rectangular prism $(V)$ = length $(l)$ × width $(w)$ × height $(h)$

$72 \text{ ft}^3 = 4 \text{ ft} \times 3 \text{ ft} \times h$

$72 \text{ ft}^3 = 12 \text{ft}^2 h$

$\dfrac{72}{12} = \dfrac{12h}{12}$

$h = 6 \text{ ft}$

Surface Area (SA) of a rectangular prism = $2lw + 2lh + 2hw$

$SA = 2(4 \text{ ft})(3 \text{ ft}) + 2(4 \text{ ft})(6 \text{ ft}) + 2(6 \text{ ft})(3 \text{ ft}) = 108 \text{ ft}^2$

The surface area of the box is $108 \text{ ft}^2$.

**Write/Explain:** I used the **Divide and Conquer** strategy. There were two parts. *First,* I found the missing dimension of the box, which was the height. I did this by using the formula for volume of a rectangular prism, and solving the equation for the height. *Second,* I used the formula for surface area of a rectangular prism.

**2. Paula Tries It. (page 96)**

**Score the Answer:** I would give Paula a **2**. She listed the question, gave all of the keywords, listed the facts except for the volume, and listed the strategy she

## 8. Measurement

used and explained how she used it. She realized that the surface area of the box was equal to total area of the six faces of the box. But Paula thought that every face had the same surface area as the base. Only one other face of the box has the same dimensions as the base.

**Make it a 4!** Rewrite.

First, Paula needs to add one more fact.

**Facts:** A box has a volume of 72 cubic feet. The base of the box has a length of 4 feet. The base of the box has a width of 3 feet.

Next, Paula can use the volume, length, width, and the formula for volume to find the height.

*Volume* of a rectangular prism $(V) =$ length $(l) \times$ width $(w) \times$ height $(h)$

$72 \text{ ft}^3 = 4 \text{ ft} \times 3 \text{ ft} \times h \rightarrow 72 \text{ ft}^3 = 12 \text{ ft}^2 h$

$\dfrac{72 \text{ ft}^3}{12 \text{ ft}^2} = \dfrac{12 \text{ ft}^2 h}{12 \text{ ft}^2 h} \rightarrow 6 \text{ ft} = h$

Now that Paula knows all three dimensions of the box, she can use a formula to find the surface area.

*Surface Area (SA)* of a rectangular prism = $2lw + 2lh + 2hw$

$SA = 2(4 \text{ ft})(3 \text{ ft}) + 2(4 \text{ ft})(6 \text{ ft}) + 2(6 \text{ ft})(3 \text{ ft}) = 108 \text{ ft}^2$

The surface area of the box is $108 \text{ ft}^2$.

I used the **Divide and Conquer** strategy. There were 2 parts. *First,* I found the height, the missing dimension of the box. I substituted the volume, length, and width of the box into the formula for volume to solve the equation for the height. *Then* I used the formula for surface area of a box. I substituted the length, width, and height into the formula, and calculated the surface area.

### Guided Problem #2

A 150-pound man would weigh about 9 pounds on Pluto. A sport-utility vehicle called the Metropolitan weighs 2.2 tons. About how many pounds would the Metropolitan weigh on Pluto?

**Keywords:** ? ?

**1. Try It Yourself.**

Answer the questions below to get a score of **4**.

What **question** are you being asked?

_____

_____

What are the **keywords**?

_____

_____

NOTICE: Photocopying any part of this book is prohibited by law.

# 8. Measurement

What are the **facts** you need to solve the problem?

_____

_____

What **strategy** can you use to solve the problem?

_____

_____

**Solve** the problem.

_____

_____

**Write/Explain** what you did to solve the problem.

_____

_____

_____

**Hint**

Possible answers include: **Logical Thinking, Divide and Conquer, Make It Simpler,** and **Write a Number Sentence or Algebraic Equation**.

**Reflect.** Review and improve your work.

_____

_____

Use the rubric on *page 13* to score this problem.

## 2. Melissa Tries It.

### Melissa's Paper

**Question:** How many pounds would the Metropolitan weigh on Pluto?

**Keywords:** pounds, tons, weigh

**Facts:** A 150-pound man would weigh about 9 pounds on Pluto. The Metropolitan weighs 2.2 tons.

**Strategy:** I used the Divide and Conquer strategy.

**Solve:**

$$\frac{\text{weight of man on Earth}}{\text{weight of man on Pluto}} = \frac{\text{weight of SUV on Earth}}{\text{weight of SUV on Pluto}}$$

$$\frac{150}{9} = \frac{2.2}{n} \rightarrow \frac{50}{3} = \frac{2.2}{n}$$

$$50n = 6.6$$

$$\frac{50n}{50} = \frac{6.6}{50} \quad n = 0.132$$

The SUV weighs 0.132 ton on Pluto.

$0.132 \text{ t} \times 2{,}000 = 26.4 \text{ lb}$

The Metropolitan would weigh about 26.4 pounds on Pluto.

**Write/Explain:** I used the Divide and Conquer strategy. There were two parts. First, I wrote and solved a proportion in which the ratios compared weight on Earth to weight on Pluto. The proportion showed the Metropolitan would weigh about 0.132 tons on Pluto. Then, when I converted that to pounds, I found that the Metropolitan would weigh about 26.4 pounds on Pluto.

# 8. Measurement

**Score the Answer.**

According to the rubric, from **1** to **3** what score would you give Melissa? Explain why you gave that score.

_____

_____

_____

**Make it a 4!** Rewrite.

_____

_____

_____

_____

## 3. Jamaal Tries It.

Remember there is often more than one way to solve a problem. Here is how Jamaal solved this problem.

### Jamaal's Paper

**Question:** How many pounds would the Metropolitan weigh on Pluto?

**Keywords:** pounds, tons, weigh

**Facts:** 150 pounds on Earth is about 9 pounds on Pluto. The Metropolitan weighs 2.2 tons.

**Strategies:** I used Logical Thinking and I Made the Problem Simpler.

$$\frac{\text{weight of man on Pluto}}{\text{weight of man on Earth}} = \frac{9}{150} = 0.06$$

To convert a weight on Earth to a weight on Pluto, multiply the weight on Earth by 0.06.

weight of Metropolitan on Earth in pounds = 
2.2 × 2,000 = 4,400

It is easier to work with 44 than 4,400. I solved the problem for a weight of 44 pounds on Earth. Then I multiplied the answer by 100.

0.06 × 44 = 2.64

2.64 × 100 = 264

The Metropolitan would weigh 264 pounds on Pluto.

**Write/Explain:** I used Logical Thinking and solved a simpler problem. First, I used the weight of the man on Earth and Pluto to find out how to convert weight on Earth to weight on Pluto. The weight on Pluto is equal to 0.06 times the weight on Earth. Next, I found the weight in pounds of the Metropolitan on Earth, which was 4,400 pounds. I made the problem simpler by multiplying 0.06 × 44. Then I multiplied that product by 100.

(cont.)

NOTICE: Photocopying any part of this book is prohibited by law.

# 8. Measurement

(Jamaal's paper cont.)

**Score:** Jamaal's solution would earn a **4** on a test. He identified the question that was asked, the keywords, and facts. He picked good strategies, explaining why and how he used them. He clearly explained the steps taken to solve the problem. His paper is perfect.

## Guided Problem #2

A 150-pound man would weigh about 9 pounds on Pluto. A sport-utility vehicle called the Metropolitan weighs 2.2 tons. About how many pounds would the Metropolitan weigh on Pluto?

Keywords:

### 4. Answers to Parts 1 and 2.

**1. Try It Yourself. (pages 99–100)**

**Question:** How many pounds would The Metropolitan weigh on Pluto?

**Keywords:** pounds, tons, weigh

**Facts:** A man who weighs 150 pounds on Earth would weigh about 9 pounds on Pluto. A sport utility vehicle called The Metropolitan weighs 2.2 tons on Earth.

**Strategy:** Divide and Conquer

**Solve:** 2.2 tons = 4,400 pounds

$$\frac{\text{weight of man on Earth}}{\text{weight of man on Pluto}} = \frac{\text{weight of SUV on Earth}}{\text{weight of SUV on Pluto}}$$

$$\frac{150}{9} = \frac{4{,}400}{x}$$

$150x = 9(4{,}400)$

$150x = 39{,}600$

$$\frac{150x}{150} = \frac{39{,}600}{150}$$

$x = 264$ lb

The Metropolitan would weigh 264 pounds on Pluto.

**Write/Explain:** I used **Divide and Conquer.** *First*, I found the weight of the Metropolitan in pounds. *Next*, I wrote and solved a proportion in which the two ratios compared weight on Earth to weight on Pluto.

**2. Melissa Tries It. (page 100)**

**Score the Answer:** I would give Melissa a **3**. She listed the keywords, the facts, the strategy she used, why and how she used it, and the question. Her explanation was clear and her work indicates that she understood the problem. However, Melissa made a computation error when she converted tons to pounds.

# 8. Measurement

**Make it a 4!** Rewrite.

Correct the computation error.
0.132 × 2,000 lb = 264 lb

The Metropolitan would weigh 264 pounds on Pluto.

## Guided Problem #3

Ned enlarges the picture so that the ratio of each side of the enlargement to the corresponding side of the original is 2:1. What is the perimeter of the enlargement? What is the area of the enlargement?

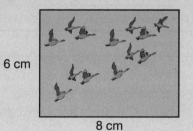

6 cm
8 cm

Keywords: ? ?

### 1. Try It Yourself.

Answer the questions to get a score of **4**.

What **question** are you being asked?

What are the **keywords**?

What are the **facts** you need to solve the problem?

What **strategy** can you use to solve the problem?

> **Hint**
> Possible answers include: **Draw a Picture, Logical Thinking,** and **Divide and Conquer.**

**Solve** the problem.

**Write/Explain** what you did to solve the problem.

**Reflect.** Review and improve your work.

NOTICE: Photocopying any part of this book is prohibited by law.

# 8. Measurement

## 2. Danny Tries It.

### Danny's Paper

**Keywords:** ratio, enlargement, perimeter, area
**Facts:** The ratio is is 2:1.
  $8 + 6 = 14$   $14 \times 2 = 28$
  $8 \times 6 = 48$   $48 \times 2 = 96$
The perimeter of the enlargement is 28 cm.
The area of the enlargement is 96 cm.

**Use the rubric on *page 13* to score his problem.**

### Score the Answer.

According to the rubric, from **1** to **3** what score would you give Danny? Explain why you gave that score.

_____
_____
_____

**Make it a 4!** Rewrite.

_____
_____
_____
_____
_____
_____

# 8. Measurement

### 3. Tara Tries It.

Remember there is often more than one way to solve a problem. Here is how Tara solved this problem.

## Tara's Paper

**Questions:** What is the perimeter of the enlargement? What is the area of the enlargement?
**Keywords:** ratio, enlargement, corresponding side, perimeter, area
**Facts:** The ratio of each side of the enlargement to the corresponding side of the original is 2:1.
The original picture is 8 cm long and 6 cm wide.
**Strategy:** Logical Thinking
First, I can find the perimeter and area of the original rectangle.

$P = 2(6 \text{ cm} + 8 \text{ cm}) = 28 \text{ cm}$
$A = 6 \text{ cm} \times 8 \text{ cm} = 48 \text{ cm}^2$

Area is the product of the sides, so doubling each side will double the perimeter.

$2 \times 28 \text{ cm} = 56 \text{ cm}$

A ratio of 2:1 means doubling the sides of the enlargement. Since perimeter is the sum of the sides, if you double the length of each side, the area will increase by 4 times, because $2 \times 2 = 4$.
$4 \times 48 \text{ cm}^2 = 192 \text{ cm}^2$

The perimeter of the enlargement is 56 centimeters. The area of the enlargement is 192 square centimeters.
**Write/Explain:** I found the perimeter and area of the original. Then I used Logical Thinking to find the perimeter and area of the enlargement. Doubling the sides doubles the perimeter and increases the area by 4 times. So, I multiplied the perimeter by 2 and the area of the original by 4 to get the perimeter and area of the enlargement.

**Score:** Tara would earn a **4** on our rubric. She identified the question that was asked, the keywords, facts, and strategy she used. She explained how she used the strategy. She explained her thinking clearly and found the correct answers. Tara labeled her work properly. Her paper is perfect!

# 8. Measurement

### 4. Answers to Parts 1 and 2.

## Guided Problem #3

Ned enlarges the picture so that the ratio of each side of the enlargement to the corresponding side of the original is 2:1. What is the perimeter of the enlargement? What is the area of the enlargement?

6 cm

8 cm

**Keywords:** ? ?

**1. Try It Yourself. (page 103)**

**Questions:** What is the perimeter of the enlargement? What is the area of the enlargement?

**Keywords:** ratio, enlargement, corresponding side, perimeter, area

**Facts:** The ratio of each side of the enlargement to the corresponding side of the original is 2:1. The diagram shows that the original picture is 8 cm by 6 cm.

**Strategies:** Logical Thinking and Divide and Conquer

**Solve:**

A ratio of 2:1 doubles each side of the original picture.

$l$(enlargement) = 2 × 8 cm = 16 cm
$w$(enlargement) = 2 × 6 cm = 12 cm

$P = 2(l + w) = 2(16 \text{ cm} + 12 \text{ cm}) = 56 \text{ cm}$

The perimeter of the enlargement is 56 centimeters.

$A$(enlargement) = $lw$ = 16 cm × 12 cm = 192 cm$^2$

The area of the enlargement is 192 square centimeters.

**Write/Explain:** I used **Logical Thinking** and the **Divide and Conquer** strategy. First, I used the ratio of the enlargement to the original to find the sides of the enlargement. Then I used the sides of the enlargement and the formulas for perimeter and area of a rectangle to find the perimeter and area of the enlargement.

# 8. Measurement

## 2. Danny Tries It. (page 104)

**Score the Answer:** I would give Danny a **1**.
Danny listed some of the keywords. He did not name the strategy he used. His listing of the facts was incomplete and his answers were incorrect.
In addition, Danny did not explain how he found his answers.

**Make it a 4!** Rewrite.

**Questions:** What is the perimeter of the enlargement? What is the area of the enlargement?

**Keywords:** ratio, enlargement, corresponding side, perimeter, area

**Facts:** The ratio of each side of the enlargement to the corresponding side of the original is 2:1. The diagram shows that the original picture is 8 cm by 6 cm.

**Strategies:** Logical Thinking and Divide and Conquer

**Solve:** I used Logical Thinking and the Divide and Conquer strategy.

A ratio of 2:1 doubles each side of the original picture.

$l$ (enlargement) = 2 × 8 cm = 16 cm
$w$ (enlargement) = 2 × 6 cm = 12 cm

$P = 2(l + w) = 2(16 \text{ cm} + 12 \text{ cm}) = 56 \text{ cm}$

The perimeter of the enlargement is 56 centimeters.

$A$(enlargement) = $lw$ = 16 cm × 12 cm = 192 cm$^2$

The area of the enlargement is 192 square centimeters.

**Write/Explain:** I used **Logical Thinking** and the **Divide and Conquer** strategy. First, I used the ratio of the enlargement to the original to find the sides of the enlargement. Then I used the sides of the enlargement and the formulas for perimeter and area of a rectangle to find the perimeter and area of the enlargement.

# 8. Measurement

# Quiz Problems

Here are some problems for you to try. Keep your rubric handy while you solve the problem. Let's see if you can score a **4**.

**1.** A circular fountain has a radius of 2 yards. Around the fountain is a circular path that is 1 yard wide. What is the area of the path?

**2.** The diagram below shows a room in which three of the corners are right angles. Tessa plans to tile the floor with Misty Brown ceramic tiles that cost $3.50 per square foot. Because of how the tile is to be cut and installed, Tessa will need to order an extra 20% of tile. How much will the tiles cost Tessa?

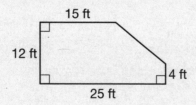

**3.** Natasha enters a gift shop and learns that sterling silver is composed of 925 parts of silver and 75 parts of copper. A sterling silver goblet weighs 0.5 pounds. How many ounces of the sterling silver goblet are silver?

**4.** A square has a perimeter of 24 inches. Each side of the original square is tripled. What is the area of the new square?

# 8. Measurement

**5.** The swimming pool below is a right triangular prism. The pool is being filled at the rate of 5 cubic feet per minute. At this rate how long will it take for the pool to be filled with water?

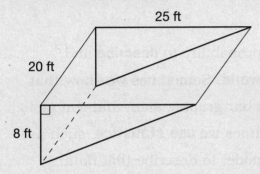

**6.** The indoor track shown below is composed of two semi-circles separated by straight track. The dimensions of the track are shown below. How many laps does a runner need to complete in order to run 1,600 meters?

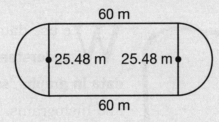

**7.** A sculptor makes a clay cube with sides of 9 inches. He cuts out an opening in the shape of a pyramid in the clay as shown below. What fraction of the original clay cube is left?

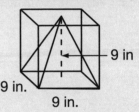

# 9. Data Analysis and Probability

We use data and probability to **describe** and **understand the world**. Sometimes we show that **data in graphs**, such as bar graphs, stem-and-leaf plots, and histograms. Sometimes we **use statistics**, such as the mean, median, or mode, to describe that data. Sometimes we talk in terms of the **probability** of an event, such as when we talk about the chance it may rain tomorrow. The method we choose to describe data depends on the situation.

# 9. Data Analysis and Probability

Many open-ended math problems deal with data, statistics, and probability. Let's look at a problem with a **score of 4**, that involves a graph and some statistics.

## Modeled Problem

This stem-and-leaf plot shows the number of Runs Batted In (RBI) that a baseball player had in each of his 12 seasons. The player's agent says that the player's typical RBI total is 105. Which **measure of central tendency** did the agent use to describe a typical RBI total, the **mean, median,** or the **mode?** Explain if one of the other measures of tendency gives a better idea of the typical RBI total. Explain your answer.

**Ned Kelly's RBI Totals**

| Stem | Leaf |
|---|---|
| 7 | 9 |
| 8 | 1 3 4 5 7 8 9 |
| 9 | 0 |
| 10 | 5 5 |
| 11 | 0 |

Key: 8 | 1 Means 81
11 | 0 Means 110

**Keywords: measure of central tendency, mean, median, mode**

### 1. Read and Think

Let's **read** the problem.

What **questions** are we asked?

- Which measure did the agent use to describe a typical RBI total, the mean, median, or mode?

Do we recognize what the **keywords** are?

- measure of central tendency, mean, median, mode

What **facts** are we given?

- The stem-and-leaf plot shows number of RBIs of a baseball player for each of his 12 seasons.

### 2. Select a Strategy

To solve this problem, we will **Use the Data From the Graph** and **Logical Thinking**.

### 3. Solve

The mean is equal to the sum of the data divided by the number of pieces of data.

Mean =

$$\frac{(79 + 81 + 83 + 84 + 85 + 87 + 88 + 89 + 90 + 105 + 105 + 110)}{12}$$

$$= \frac{1,086}{12} = 90.5$$

# 9. Data Analysis and Probability

The median is the middle number in a set of data when the numbers are arranged in order. Since there are 12 numbers in this set, the median is the mean of the 6th and 7th numbers.

Median = $\frac{87 + 88}{2}$ = 87.5

The mode is the number that occurs the most. Since 105 is the only number that occurs more than once, it is the mode. Let's decide which measure best describes the data. Since the mode of 105 is much greater than 9 of the 12 pieces of data, it does not provide a good idea of the typical RBI total. The mean of 90.5 is closer to the rest of the data than was the mode. But it too is greater than 9 of the data points. So, the median probably gives the best idea of the typical RBI total.

## 4. Write/Explain

We found the mean, median, and mode, and saw that the agent used the mode to describe a typical RBI total. Then we used **Logical Thinking**. We compared each measure of central tendency to the data, and decided that the mean and the median are both better ways of describing a typical RBI total.

## 5. Reflect

Let's review our work and answer.

- Did we show that we knew what the problem asked? **Yes.**
- Did we know what the keywords were? **Yes.**
- Did we show that we knew what facts were given? **Yes.**
- Did we name and use the correct strategy? **Yes.**
- Was our math correct? **Yes. We checked it. It was correct.**
- Did we label our work? **Yes.**
- Was our answer correct? **Yes.**
- Did we explain why we chose the strategy and how it was used to get to the right answer? **Yes.**
- Were all of our steps included? **Yes.**
- Did we write a good, clear explanation of our work? **Yes.**

This solution would earn a **4** on our rubric. We got the correct answer and gave a clear explanation of how we arrived at the answer.

On the following pages are some **Guided Open-Ended Math Problems.** For each problem there are **four parts**. In the **first part**, you will solve the problem with guided help. In the **second part**, you will score and correct a solution with guided help. The **third part** shows one solution that scores a perfect **4**. This solution may or may not differ from your way. The **fourth part** has *answers* to the **first** and **second parts** so you can check your work.

# 9. Data Analysis and Probability

## Guided Problem #1

Lena has a bag with 3 red tiles, 4 blue tiles, and 2 green tiles. She is going to pick a tile without looking and keep the tile. Then she is going to pick a second tile without looking. Lena says that the probability of picking 2 blue tiles is exactly twice as great as the probability of picking 2 green tiles. Is Lena correct? Explain your answer.

Keywords:

### 1. Try It Yourself.

Answer the questions below to get a score of **4**.

What **question** are you being asked?

_____

_____

What are the **keywords**?

_____

_____

_____

What are the **facts** you need to solve the problem?

_____

_____

What **strategy** can you use to solve the problem?

_____

_____

**Solve** the problem.

_____

_____

**Hint**

Possible answers include: **Logical Thinking, Act it Out,** and **Divide and Conquer.**

**Write/Explain** what you did to solve the problem.

_____

_____

_____

_____

**Reflect.** Review and improve your work.

_____

_____

_____

Use the rubric on *page 13* to score this work.

NOTICE: Photocopying any part of this book is prohibited by law.

# 9. Data Analysis and Probability

## 2. Trent Tries It.

### Trent's Paper

**Question:** Lena says that the probability of picking 2 blue tiles is exactly twice as great as the probability of picking 2 green tiles. Is Lena correct?

**Keywords:** probability, twice, as great

**Facts:** There are 4 blue tiles. There are 2 green tiles. Lena will pick a tile without looking. She will keep the tile. She will pick another tile without looking.

**Strategy:** Logical Thinking

**Solve:**
P(blue, given blue) = $\frac{4}{6} \times \frac{3}{5} = \frac{2}{5}$

P(green, given green) = $\frac{2}{6} \times \frac{1}{5} = \frac{1}{15}$

Lena is wrong. The probability of picking 2 blue tiles is more than twice as great as the probability of picking 2 green tiles.

**Write/Explain:** I used Logical Thinking. I found the probability of the first pick and the probability of the second pick. Then I multiplied. I did that for picking 2 blue tiles. Then I did it for picking 2 green tiles. Since $\frac{2}{5}$ is more than 2 times $\frac{1}{15}$, Lena was wrong.

**Score the Answer.**

According to the rubric, from **1** to **3** what score would you give Trent? Explain why you gave that score.

_____
_____
_____
_____

**Make it a 4!** Rewrite.

_____
_____
_____
_____
_____
_____
_____
_____

Use the rubric on *page 13* to score this work.

NOTICE: Photocopying any part of this book is prohibited by law.

# 9. Data Analysis and Probability

## 3. Kara Tries It.

Remember there is often more than one way to solve a problem. Here is how Kara solved this problem.

### Kara's Paper

**Question:** Lena says that the probability of picking 2 blue tiles is exactly twice as great as the probability of picking 2 green tiles. Is Lena correct?

**Keywords:** probability, twice, as great

**Facts:** A bag has 3 red tiles, 4 blue tiles, and 2 green tiles. Lena will pick a tile without looking, keep the tile, and pick a second tile without looking.

**Strategies:** Act It Out and Write a Number Sentence

**Solve:** I can cut out paper squares and use them to show the tiles: R = red, B = blue, G = green.

The probability of picking B on the first pick is $\frac{4}{9}$.

When I take one B away, the probability of picking another B is $\frac{3}{8}$.

So, the probability of picking two blue tiles is: $\frac{4}{9} \times \frac{3}{8} = \frac{12}{72} = \frac{1}{6}$

The probability of picking G on the first pick is $\frac{2}{9}$.

When I take one G away, the probability of picking another G is $\frac{1}{8}$.

So, the probability of picking two green tiles is:
$\frac{2}{9} \times \frac{1}{8} = \frac{2}{72} = \frac{1}{36}$

Twice the probability of picking 2 green tiles =
$\frac{1}{36} \times 2 = \frac{2}{36} = \frac{1}{18}$

$\frac{1}{6} \neq \frac{1}{18}$, so Lena is incorrect.

**Write/Explain:** I Acted It Out. I used 9 pieces of paper to show the 9 tiles. I multiplied the two probabilities to find the probability of picking blue on both picks. I did the same thing for picking green. My calculations showed that picking the probability of picking 2 blue tiles was not exactly twice as great as the probability of picking 2 green tiles. So, Lena was wrong.

**Score:** Kara's solution would earn a **4** on a test. Kara identified the the question that was asked, the keywords, the facts, and picked a good strategy. She explained how she used the strategy. She clearly explained the steps taken to solve the problem. Kara labeled her work properly. It is perfect!

# 9. Data Analysis and Probability

## 4. Answers to Parts 1 and 2.

### Guided Problem #1

Lena has a bag with 3 red tiles, 4 blue tiles, and 2 green tiles. She is going to pick a tile without looking and keep the tile. Then she is going to pick a second tile without looking. Lena says that the probability of picking 2 blue tiles is exactly twice as great as the probability of picking 2 green tiles. Is Lena correct? Explain your answer.

**Keywords:** ??

**1. Try It Yourself. (page 113)**

**Question:** Lena says that the probability of picking 2 blue tiles is exactly twice as great as the probability of picking 2 green tiles. Is Lena correct?

**Keywords:** probability, twice, as great

**Facts:** A bag has 3 red tiles, 4 blue tiles, and 2 green tiles. Lena will pick a tile without looking, keep the tile, and pick a second tile without looking.

**Strategy:** Logical Thinking

**Solve:**

First, find the probability of picking 2 blue tiles.

1st pick: 4 of 9 tiles are blue. $P(\text{blue}) = \frac{4}{9}$
2nd pick: 3 of 8 tiles are blue. $P(\text{blue, given blue}) = \frac{3}{8}$
$P(\text{2 blue tiles}) = \frac{4}{9} \times \frac{3}{8} = \frac{12}{72} = \frac{1}{6}$

Next, find the probability of picking 2 green tiles.

1st pick: 2 of 9 tiles are green. $P(\text{green}) = \frac{2}{9}$
2nd pick: 1 of 8 tiles are green. $P(\text{green, given green}) = \frac{1}{8}$
$P(\text{2 red tiles}) = \frac{2}{9} \times \frac{1}{8} = \frac{2}{72} = \frac{1}{36}$

Double the probability of picking 2 green tiles:

$2 \times \frac{1}{36} = \frac{2}{36} = \frac{1}{18}$. Since $\frac{1}{18} \neq \frac{1}{6}$,

Lena is incorrect.

**Write/Explain:** I used **Logical Thinking**. I found the probability of picking 2 blue tiles and the probability of picking 2 green tiles. I then doubled the probability of picking 2 green tiles to see if it would match the probability of picking 2 blue tiles. It did not, so Lena was incorrect.

# 9. Data Analysis and Probability

## 2. Trent Tries It. (page 114)

> **Score:** I would give Trent a 2. He listed the keywords, the strategy he used, including how he used it, and the question, but he left out a fact: 3 red tiles. Trent's answer shows that he understood how to find the probability for a pair of dependent events, such as P(blue, given blue) and P(green, given green). But Trent ignored the fact that the bag included red tiles. That caused him to find incorrect probabilities. Trent also didn't give a full explanation of his solution process. Although Trent found that Lena was incorrect, the way he arrived at that answer was incomplete and partly incorrect.

**Make it a 4!** Rewrite.

First, Trent needs to add a fact.

**Facts:** There are 3 red tiles. 4 blue tiles. There are 2 green tiles. Lena will pick a tile without looking. She will keep the tile. She will pick another tile without looking.

Next, Trent needs to recalculate the probabilities, based on the bag, starting with 9 tiles instead of 6 tiles.

First, find the probability of picking 2 blue tiles.

1st pick: 4 of 9 tiles are blue. $P(\text{blue}) = \frac{4}{9}$
2nd pick: 3 of 8 tiles are blue. $P(\text{blue, given blue}) = \frac{3}{8}$
$P(\text{2 blue tiles}) = \frac{4}{9} \times \frac{3}{8} = \frac{12}{72} = \frac{1}{6}$

Next, find the probability of picking 2 green tiles.

1st pick: 2 of 9 tiles are green. $P(\text{green}) = \frac{2}{9}$
2nd pick: 1 of 8 tiles are green. $P(\text{green, given green}) = \frac{1}{8}$
$P(\text{2 red tiles}) = \frac{2}{9} \times \frac{1}{8} = \frac{2}{72} = \frac{1}{36}$

Double the probability of picking 2 green tiles:
$2 \times \frac{1}{36} = \frac{2}{36} = \frac{1}{18}$. Since $\frac{1}{18} \neq \frac{1}{6}$,

Lena is incorrect.

**Write/Explain:** I used **Logical Thinking**. I found the probability of picking 2 blue tiles and the probability of picking 2 green tiles. I then doubled the probability of picking 2 green tiles to see if it would match the probability of picking 2 blue tiles. It did not, so Lena was incorrect.

# 9. Data Analysis and Probability

## Guided Problem #2

The scatterplot shows the number of pages of a book read by a student in given amounts of time. About how many pages do you think the student will read in 4 hours?

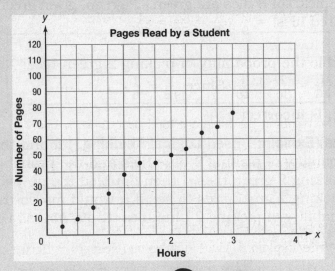

Keywords: ? ?

### 1. Try It Yourself.

Answer the questions below to get a score of **4**.

What **question** are you being asked?

_____

What are the **keywords**?

_____

What are the **facts** you need to solve the problem?

_____

What **strategy** can you use to solve the problem?

_____

**Solve** the problem.

_____

_____

**Write/Explain** what you did to solve the problem.

_____

_____

**Reflect.** Review and improve your work.

_____

### 2. Fran Tries It.

#### Fran's Paper

Question: About how many pages do you think the student will read in 4 hours?
Keywords: scatterplot, pages, hours
The student will read 120 pages in 4 hours.

NOTICE: Photocopying any part of this book is prohibited by law.

# 9. Data Analysis and Probability

## Sandra's Paper

**Question:** About how many pages do you think the student will read in 4 hours?

**Keywords:** scatterplot, pages, hours

**Facts:** The points on the scatterplot show the number of pages read in given amounts of time.

**Strategies:** Make a Table and Look For a Pattern.

| Hours | 1 | 2 | 3 | 4 |
|---|---|---|---|---|
| Pages Read | about 26 or 27 | 50 | about 75 | ? |

The points in the table are for 1 hour, 2 hours, and 3 hours.

The points come close to making a pattern: 25 pages, 50 pages, 75 pages. In this pattern, the number of pages increases about 25 pages each hour.

The pattern gives 100 pages for 4 hours.

**Write/Explain:** I Made a Table and Found a Pattern. I needed to find the point for 4 hours. So, I chose just 3 of the points from the scatterplot: 1 hour, 2 hours, and 3 hours. I found the pattern for the first 3 hours and extended it to 4 hours. Since the pattern showed that the student read about 25 pages an hour, the student would read about 100 pages in 4 hours. I believe that choosing just three points to find the pattern works, because all the points on the graph are close to forming a line. There are no points that would fall much above or below that line. (cont.)

Use the rubric on *page 13* to score this problem.

**Score the Answer.**

According to the rubric, from **1** to **3** what score would you give Fran? Explain why you gave that score.

_____
_____
_____

**Make it a 4!** Rewrite.

_____
_____
_____
_____

### 3. Sandra Tries It.

Remember there is often more than one way to solve a problem. Here is how Sandra solved this problem.

NOTICE: Photocopying any part of this book is prohibited by law.

# 9. Data Analysis and Probability

(Sandra's paper cont.)

**Score:** Sandra's solution would earn a **4** on a test. She identified the question that was asked, the keyword, the facts, and picked two good strategies. She explained why and how she used these strategies. She clearly explained the steps taken to solve the problem. Sandra labeled her work. Sandra's solution is perfect!

## 4. Answers to Parts 1 and 2.

## Guided Problem #2

The scatterplot shows the shows the number of pages of a history book read by a student in given amounts of time. Based on the scatterplot, about how many pages do you think the student will read in 4 hours?

**Keywords:** ? ?

### 1. Try It Yourself. (page 118)

**Question:** About how many pages do you think the student will read in 4 hours?

**Keywords:** scatterplot, pages, hours

**Facts:** The points on the scatterplot show the number of pages read in given amounts of time.

**Strategy:** Draw a Graph.

NOTICE: Photocopying any part of this book is prohibited by law.

# 9. Data Analysis and Probability

**Solve:**

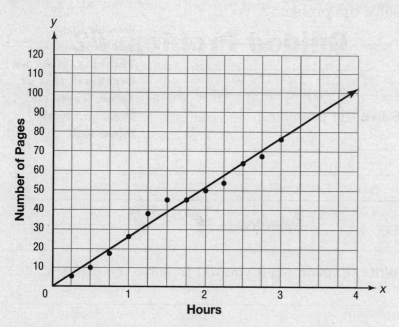

The line shows that the student will read about 100 pages in 4 hours.

**Write/Explain: I Drew a Graph.** Then, I drew the line that came closest to connecting the points on the graph. The point on the line for 4 hours is (4, 100). So, line shows that the student will read about 100 pages in 4 hours.

## 2. Fran Tries It. (pages 118–119)

**Score the Answer:** I would give Fran a **1**. Fran listed the keywords and the question, but did not list any facts. Fran gave an answer, which was close, but was incorrect. She did not list a strategy or explain how she found the answer.

**Make it a 4!** Rewrite.

**Facts:** The points on the scatterplot show the number of pages read in given amounts of time.

**Solve: I Drew a Graph.**

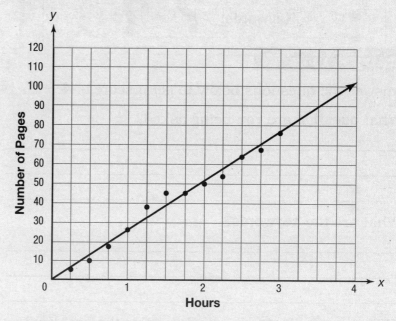

The line shows that the student will read about 100 pages in 4 hours.

I drew the line that came closest to connecting the points on the graph. The point on the line for 4 hours is (4, 100). So, the line shows that the student will read about 100 pages in 4 hours.

# 9. Data Analysis and Probability

## Guided Problem #3

A set of five numbers has a mean of 8, a median of 9, a mode of 12, and a range of 10. What are the five numbers?

Keywords: ??

### 1. Try It Yourself.

Answer the questions below to get a score of **4**.

What **question** are you being asked?

_____
_____
_____

What are the **keywords**?

_____
_____
_____

What are the **facts** you need to solve the problem?

_____
_____
_____

What **strategy** can you use to solve the problem?

_____
_____
_____

**Solve** the problem.

_____
_____
_____
_____
_____
_____
_____

**Write/Explain** what you did to solve the problem.

_____
_____
_____
_____
_____

**Reflect.** Review and improve your work.

_____
_____
_____
_____
_____

> **Hint**
>
> Possible answers include: **Logical Thinking, Make an Organized List or Table, Guess and Test,** and **Work Backward.**

NOTICE: Photocopying any part of this book is prohibited by law.

# 9. Data Analysis and Probability

## 2. Ben Tries It.

### Ben's Paper

**Question:** What are the 5 numbers?

**Keywords:** mean, median, mode, range

**Facts:** There are 5 numbers with a mean of 8, a median of 9, a mode of 12, and a range of 10.

**Strategy:** Logical Thinking

**Solve:**

The mode is 12. So, 12 happens at least twice.   12   12

The median is 9. That means 9 is the middle number

9   12   12

Since there are 5 numbers, the other 2 will be less than 9. The mean is 8. That means the 5 numbers must have a total of 5 × 8 = 40. We already know three of the numbers. Subtract them from 40.

40 − (9 + 12 + 12) = 7.

7 is left for the last two numbers. I can use 3 and 4. So, the numbers are 3, 4, 9, 12, and 12

**Write/Explain:** I used Logical Thinking. I used what I knew about the mode and the median to find three out of the five numbers. Then I used the mean to find the other two numbers.

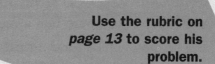

Use the rubric on *page 13* to score his problem.

**Score the Answer.**

According to the rubric, from **1** to **3** what score would you give Ben? Explain why you gave that score.

_____
_____
_____
_____

**Make it a 4!** Rewrite.

_____
_____
_____
_____

**Reflect.** Review and improve your work.

_____
_____
_____
_____

NOTICE: Photocopying any part of this book is prohibited by law.

# 9. Data Analysis and Probability

## 3. Alma Tries It.

Remember there is often more than one way to solve a problem. Here is how Alma solved this problem.

### Alma's Paper

**Question:** What set of 5 numbers meet the conditions from the facts?

**Keywords:** mean, median, mode, range

**Facts:** There are 5 numbers. They have a mean of 8, a median of 9, a mode of 12, and a range of 10.

**Strategies:** Make an Organized List and Guess and Test. There are five numbers.

____ , ____ , ____ , ____ , ____

The mode is 12. Let's say 12s are 3 out of the 5 numbers. If the mode of 12 happens 3 times, it will also be the middle number. That makes it the median.

____ , ____ , 12, 12, 12,  or  ____ , 12, 12, 12, ____  or  12, 12, 12, ____ , ____ .

But 9 is the median. So 12 must happen only twice.

____ , ____ , 9, 12, 12

The range of the data is 10. 12 − 10 = 2

2, ____ , 9, 12, 12

Find a fifth number that will give a mean of 8. Try 4.

2, 4, 9, 12, 12

$$\frac{(2 + 4 + 9 + 12 + 12)}{5} = 7.8$$

4 was too low. Try 5. That works.

2, 5, 9, 12, 12

$$\frac{(2 + 5 + 9 + 12 + 12)}{5} = 8$$

The 5 numbers are 2, 5, 9, 12, 12

**Write/Explain:** I Made an Organized List and used the Guess and Test strategy. I used the list, the mode, and the median to find the three greatest numbers in the set of data. I used the range to find the least number. Then I used the mean to guess the other number. It took me two guesses to find the final number in the set.

**Score:** Alma's solution would earn a 4 on our rubric. She identified the question that was asked, the keywords, the facts, and picked two good strategies. She explained why and how she used the strategies. Then she clearly explained the steps she took in order to solve the problem. Alma labeled her work. She did a great job!

# 9. Data Analysis and Probability

### 4. Answers to Parts 1 and 2.

## Guided Problem #3

A set of five numbers has a mean of 8, a median of 9, a mode of 12, and a range of 10. What are the five numbers?

Keywords: ? ?

### 1. Try It Yourself. (page 122)

**Questions:** What are the 5 numbers?

**Keywords:** mean, median, mode, range

**Facts:** A set of five numbers has a mean of 8, a median of 9, a mode of 12, and a range of 10.

**Strategy:** Logical Thinking

**Solve:** The mode is 12. That means 12 happens at least twice.
12, 12

The median is 9. That means 9 is the middle number. Since there are 5 numbers, the other 2 will be less than 9.
?, ?, 9, 12, 12

The range is 10. 12 − 10 = 2. The least number is 2.
2, ?, 9, 12, 12

The mean is 8. Five numbers with a mean of 8 add to 40. So, find a fifth number that gives the set a sum of 40.

$5 \times 8 = 40$
$2 + ? + 9 + 12 + 12 = 40$
$? + 35 = 40$

The fifth number is 5.

The numbers are 2, 5, 9, 12, and 12.
? = 5

**Write/Explain:** I used **Logical Thinking**. I used the mode and the median to find that three of the numbers were 9, 12, and 12. I also found that the other 2 numbers must be less than 9. I used the range to find that the least number must be 2. Then I used the mean to find the last number in the set.

### 2. Ben Tries It. (page 123)

**Score the Answer:** I would give Ben a **3**. He listed the question, keywords, and the facts. Ben chose a good strategy, explaining how he used it. He clearly explained the steps that he took. However, Ben forgot that the set of numbers has a range of 10. As a result, Ben got the wrong answer.

**Make it a 4!** Rewrite.

Use the range to find the least number. 12 − 10 = 2.

Now the set of numbers is 2, ?, 9, 12, 12.

The mean is 8. That means the five numbers must have a total of $5 \times 8 = 40$. We already know four of the numbers. Subtract them from 40.

$40 − (2 + 9 + 12 + 12) = 5$

The numbers are 2, 5, 9, 12, 12.

NOTICE: Photocopying any part of this book is prohibited by law.

# 9. Data Analysis and Probability

# Quiz Problems

Here are some problems for you to try. Keep your rubric handy while you solve the problem. Let's see if you can score a **4**.

**1.** A rectangle is 15 centimeters by 20 centimeters. Inside of that rectangle is a right triangle with legs of 6 centimeters and 5 centimeters. A dart is thrown at the rectangle. Suppose the dart is equally likely to hit any point in the rectangle. What is the probability that the dart will hit the right triangle?

**2.** The box-and-whisker plot below shows the ages of members of the Metropolitan Chess Club. The club has 36 members. No two of the members are the same age. How many of the members are over 56 years old?

**3.** A basketball player makes 65% of his free throws. How can you use a number cube to simulate the results of the basketball player's next 10 free throws? How close does your method come to simulating a 65% rate of making free throws?

**4.** The ages of teachers at Midland Middle School are shown below.
27, 28, 34, 48, 52, 61,
54, 37, 60, 36, 40, 48,
52, 34, 49, 49, 44, 32,
33, 44, 55, 29, 42, 48,
56, 29, 38, 29, 46, 47
a. Construct a graph that shows the data in intervals.
b. Is there another graph that you could construct that would show the data in intervals? Would one of the graphs show information that the other does not?

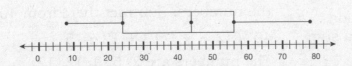

# 9. Data Analysis and Probability

**5.** The histogram shows the distances that Ned hits golf balls on the driving range. Ned considers drives of less than 200 yards to be unacceptable and drives of 200 yards or more to be acceptable. Based on the histogram, what is the probability that exactly one of Ned's next two drives will be acceptable and the other drive will be unacceptable?

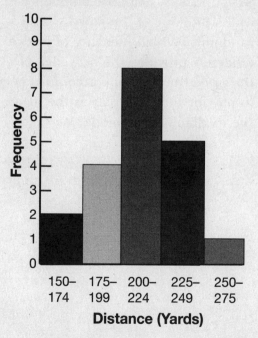

**6.** Inez runs in a 5-mile race to raise money for medical research. Her 10 sponsors give her the following donations: $7.50, $5, $7.50, $100, $10, $5, $10, $5, $15, $8. Which measure of central tendency would give the most accurate idea of a typical donation, the mean, the median, or the mode?

**7.** The scatterplot shows data for apartment sizes and monthly rents in a neighborhood.

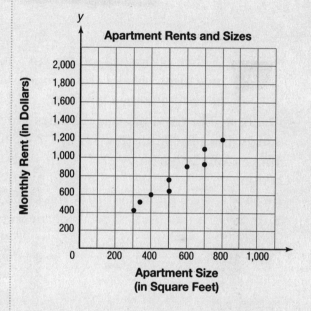

Based on the scatterplot, about how much would you expect the monthly rent to be for an apartment that is 1,000 square feet?

# 10. Test #1

**Answer the following questions to the best of your ability. Remember, even if you are unsure of how to solve the problem, you will always earn some credit if you begin the problem. Good luck!**

**1.** A rope 64 inches long is cut into three pieces. The second piece is 4 inches longer than the first. The third piece is 6 inches longer than the first. How many inches long is each piece?

**2.** Deborah and Cormac are building a sandcastle. They plan to construct a rectangular pyramid with dimensions 5 feet wide by 5 feet long by 3 feet high. To transport this volume of sand, they use a cylindrical can with a diameter of 1 foot and a height of 9 inches. About how many cans of sand do they need to create this pyramid?

**3.** Nedra compares the price of the same DVD player in eight local stores. The prices are $99, $79, $89, $79, $79, $109, $79, and $85. Suppose a shopper walks into one of the stores randomly. Which measure of central tendency provides the best idea of the price that a shopper could expect to pay for the DVD player, the mean, the median, or the mode?

## 10. Test #1

**4.** In a quilting club, the students will piece together a red and green quilt using 1-foot square blocks. There will be nine equally sized pieces in each block so that the ratio of red to green will be 1:2. For example, a sample block might look like this:

| R | G | G |
|---|---|---|
| G | R | G |
| G | G | R |

If the complete quilt will be 5 blocks by 4 blocks, how many red pieces will be needed for the quilt?

**5.** A graphic artist is designing a logo for a company. She has drawn the figure shown below. Her plan for the logo includes a reflection of this shape over the y-axis. What are the coordinates of the vertices of the reflection?

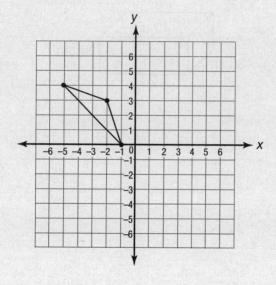

**6.** Tawana's grandmother gives her a gift of $30. After that, Tawana begins to save money. The graph below shows the total amount of money Tawana has saved.

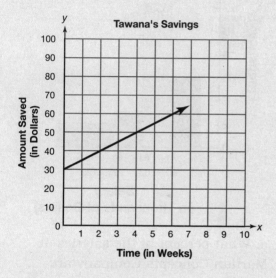

If Tawana continues to save money at the same rate, when will she have $100?

# 10. Test #1

Answer the following questions to the best of your ability. Remember, even if you are unsure of how to solve the problem, you will always earn some credit if you begin the problem. Good luck!

**7.** The histogram shows data about salaries at Martian Concepts Company.

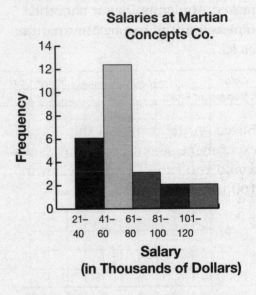

a. What percent of the salaries at Martian Concepts Company are $80,000 or less?

b. Could you use this histogram to make a stem-and-leaf plot about salaries at Martian Concepts Company? Explain.

**8.** Kendra is using some felt to line the inside of the box shown below. The thickness of the wood is negligible. What is the minimum amount of felt that Kendra needs in order to line the inside of the box (including the underside of the lid)?

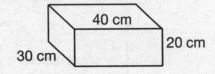

**9.** Families In Community Service (FICS) sponsored a citywide "Fun Fair" and raised $32,240 to support various city programs. They plan to give $\frac{1}{4}$ of the total to the library for literacy programs and $\frac{3}{8}$ of the total to a local chapter of Job Corp for youth. How much is left for other projects?

## 10. Test #1

**10.** From his home, Tyler biked 3 kilometers west to the pharmacy. From there, he rode 1.5 kilometers south to the post office. Then he rode 1.5 kilometers east to the library. After returning his books, Tyler rode home using the shortest possible route. To the nearest tenth of a kilometer, how far was his ride home from the library?

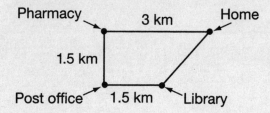

**11.** Sister's Pizzeria offers the following pizza pies on its menu.

| Medium | 12" diameter | $ 7.99 |
| Large | 14" diameter | $ 9.99 |
| Extra Large | 16" diameter | $11.59 |

Which size gives the best value?

**12.** The table shows the results of a probability experiment in which a counter was picked randomly from a bag, then replaced in the bag before the next pick.

| Color | Red | Blue | Yellow | Green |
| --- | --- | --- | --- | --- |
| Number of Picks | 11 | 3 | 7 | 9 |

Based on the results of this experiment, about how many times would you expect to pick yellow in 100 tries?

# 10. Test #1

Answer the following questions to the best of your ability. Remember, even if you are unsure of how to solve the problem, you will always earn some credit if you begin the problem. Good luck!

**13.** Toni can paint a room in 4 hours. Her sister Diana can paint the same room in 6 hours. If they work together, how long will it take them to paint the room?

**14.** How many squares of all sizes are in the figure?

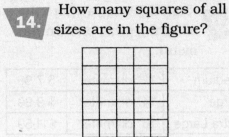

**15.** During a sale in which all books are 35% off, a book costs $15.60. What was the original price of the book before the discount?

## 10. Test #1

**16.** A target consists of a small square with sides of 5 inches that is centered within a larger square with sides of 15 inches. A dart is thrown at the target. Suppose the dart is equally likely to hit any point in the larger square. Is the probability of hitting the small square the same as the ratio that compares one side of the small square to one side of the larger square?

**17.** Kathie has 16 coins. One is a quarter and the rest are dimes and nickels. The total value of the coins is $1.55. How many dimes and how many nickels does she have?

**18.** A cylindrical soup can has the dimensions shown below. A label covers the lateral surface, which is the curved surface of the can. What is the least possible area that the label could have in order to completely cover the curved surface?

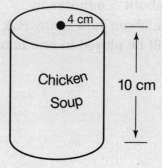

NOTICE: Photocopying any part of this book is prohibited by law.

# 10. Test #1

Answer the following questions to the best of your ability. Remember, even if you are unsure of how to solve the problem, you will always earn some credit if you begin the problem. Good luck!

**19.** Keyshaun is a DJ for his school dance. He has downloaded 500 songs onto his computer and will play these songs randomly at the dance. His collection includes 200 R & B songs, 200 Rock/Pop songs, and 100 Classic Rock songs. If each song is about 3 minutes in duration, about how many R & B songs will be played in one hour?

**20.** The diagram below shows a square window that is composed of a clear circular pane of glass and 4 tinted panes of glass. The radius of the circle is 5 inches. What is total area of the tinted glass?

# 10. Test #2

Answer the following questions to the best of your ability. Remember, even if you are unsure of how to solve the problem, you will always earn some credit if you begin the problem. Good luck!

**1.** An electronics store regularly gives a 5% discount (before tax) on all purchases of over $100. The state sales tax is 5%. Marina says that the discount means that a computer that has a price of $630 will have a total cost of $630, including the state sales tax. Is Marina correct? Explain.

**2.** A captain of a ship uses a grid in which coordinates represent kilometers. The ship travels from (3, 1) to (11, 7) to (−5, 7). From (−5, 7), the ship returns to its starting location using the most direct route. How far did the ship travel in all?

**3.** This year Nick is $\frac{1}{2}$ as old as Steve. Six years from now, Nick will be $\frac{5}{8}$ as old as Steve. At that time, the sum of their ages will be 39. How old are Nick and Steve now?

# 10. Test #2

Answer the following questions to the best of your ability. Remember, even if you are unsure of how to solve the problem, you will always earn some credit if you begin the problem. Good luck!

**4.** A radar records the speeds of drivers on one road during the hour. The results are shown below.

52, 54, 60, 65, 48, 62, 55, 49
56, 58, 47, 72, 50, 65, 64, 60

a. Construct a graph that makes it easy to identify the median of the data. What is the median?

b. Is there another graph that you could construct that would make it easy to identify the median?

**5.** Raj has a photo that is 4 inches by 6 inches.

He would like to enlarge it so that the longer side of the enlargement is 9 inches. Raj can do this on a color copier if he knows the percent of enlargement to use, which is a percent that the area is increased. For example, if Raj enlarges a picture 125%, this means that the area of the copy will be 125% of the original. What percent of enlargement should Raj select to make in enlargement in which the longer side is 9 inches?

**6.** Two cousins, Vonda and Irene, live 35 miles apart along the same road. They leave their houses at the same time and bicycle to meet each other. If Vonda can travel 8 miles per hour and Irene can travel 6 miles per hour, how long will it take for them to meet? How many miles will each of them have traveled?

## 10. Test #2

**7.** The short track for speed skating shown below consists of two line segments and two semi-circles on either end. On this track, a skater completes 45 laps in a 5,000-meter race. To the nearest meter, what is the width of the rectangular portion of the track?

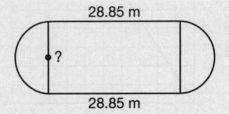

**8.** Maggie is preparing for the lacrosse season. She goes to the Sports Shop to purchase a lacrosse stick ticketed at $84 and a pair of cleats selling for $59. She has two coupons for the sports shop:

Each coupon can be used with one item. How should Maggie use the coupons to save the most money on her purchases? What is the lowest combined price she can get by using the coupons?

**9.** Jin's mother has errands to run. She needs to go to the library (L), supermarket (S), post office (PO), bookstore (B), dry cleaning store (DC), and car wash (CW). She starts from her home and the order in which she runs the errands doesn't matter. The diagram below shows the distances between these points in miles.

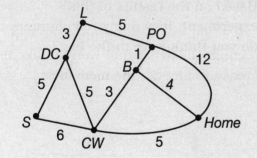

a. Is it possible for Jin's mother to drive to each place she needs to go to without driving over the any part of her route twice? If so, explain how can she do it.

b. If it is possible, what is the distance of the shortest way for Jin's mother to complete her chores, not counting the distance from her last stop back to her home?

# 10. Test #2

Answer the following questions to the best of your ability. Remember, even if you are unsure of how to solve the problem, you will always earn some credit if you begin the problem. Good luck!

**10.** The table shows the results of a probability experiment in which a counter was picked randomly from a bag, then replaced in the bag before the next pick. There were 12 counters in the bag.

| Color | Red | Blue | Yellow | Green |
|---|---|---|---|---|
| Number of Picks | 19 | 53 | 2 | 6 |

Based on the results of this experiment, how many red counters do you think were in the bag?

**11.** Yelena is baking apple squares in a rectangular pan that is 24 centimeters by 30 centimeters. She wants to make the squares as large as possible. What dimensions should she make the squares? How many squares will she have?

**12.** The graph below shows Frank's earnings from gardening.

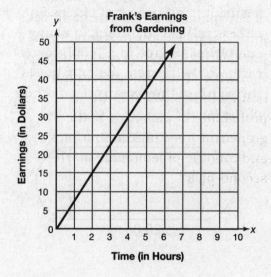

Based on the graph, how much would Frank earn in 10 hours?

138

## 10. Test #2

**13.** A cooler has 4 bottles of orange juice, 3 bottles of grapefruit juice, 2 bottles of lemonade, and 1 bottle of apple juice. Suppose you pick a bottle of juice without looking and you keep the bottle. Then you pick another bottle of juice. What is the probability of picking a bottle of grapefruit juice on the first pick, and a bottle of lemonade on the second pick?

**14.** In a parallelogram, one angle has 3 times the angle measure of another angle. What are the measures of all four angles?

**15.** The Begg family is installing an in-ground pool. A backhoe has dug out a volume of 37.5 cubic meters of dirt. The opening in the ground is in the shape of a rectangular prism with a length of 6 meters and a width of 5 meters. If the Beggs plan to line the sides and bottom of the pool with a plastic liner, how much liner do they need?

NOTICE: Photocopying any part of this book is prohibited by law.

# 10. Test #2

Answer the following questions to the best of your ability. Remember, even if you are unsure of how to solve the problem, you will always earn some credit if you begin the problem. Good luck!

**16.** In a science experiment, three sticks of different lengths were positioned vertically in the ground and their shadows were measured. The results are shown in the table below.

| | Stick A | Stick B | Stick C |
|---|---|---|---|
| Length of Stick | 12 in. | 18 in. | 30 in. |
| Length of Shadow | 14 in. | 21 in. | 35 in. |

At the same time, a tree nearby cast a shadow that was 38.5 feet long. What was the height of the tree?

**17.** There are 92 students in an assembly. The number of seventh graders is 7 more than the number of sixth graders. The number of eighth graders is 3 less than the number of seventh graders. How many students from each grade are in the assembly?

**18.** Mars is about $1.4 \times 10^8$ miles from the Sun.
Neptune is about 20 times farther from the Sun than Mars is. About how far away from the Sun is Neptune? Is Neptune more than 20 billion miles from the Sun?

# 10. Test #2

**19.** A rectangle has a perimeter of 22 inches. Each side of the original rectangle is multiplied by 4 to make a new rectangle. What is the perimeter of the new rectangle?

**20.** The list below shows the point totals for each game that the Townsend High School basketball team has played so far this season.

56, 48, 63, 55, 64, 66, 70, 64, 66, 50, 47, 58, 50

What is the minimum score needed in the next game in order to make the median greater than 60 points?

Mathematics Open-Ended Questions, Level G

# 11. Home-School Connection

Working on these questions at home with a family member is fun! Find a comfortable place to work and have all the tools you need. Discuss the problem and any strategies you might be able to use to solve the problem. Then go for it! Try to find the answer. Remember to write a clear explanation of your solution process. Don't forget to use your rubric to score your results!

# 11. Home-School Connection

Dear Family Member:

This year your child will be learning about **open-ended math questions** in mathematics class. An open-ended math question is a mathematics word problem that has one single correct answer, but that can be solved in several different ways. Open-ended math questions are extremely important on tests your child will take.

You can help your child practice solving these questions by working together on the take–home sheets in this chapter. Don't forget to use the rubric as a guide. Remember, when you work with your child, don't do the problem. Rather, encourage your child to ask questions that will lead him or her to the answer. And don't be surprised if your child arrives at the answer to the problem using a method different from the one you are used to using.

It is important that your child make his or her thinking clear to the reader. After your child has solved the problem (or gone as far as he or she can), help your child write a clear explanation of what he or she did to solve it, and why he or she decided to do it that way. This will help your child clarify his or her own thoughts.

The problems on the following pages are based on the areas of mathematics considered important in solving open-ended math problems. These are:

- **Number and Operations**
- **Algebra**
- **Geometry**
- **Measurement**
- **Data Analysis and Probability**

Enjoy!

NOTICE: Photocopying any part of this book is prohibited by law.

# 11. Home-School Connection

# Number and Operations

**Problem**

Some seventh-grade students sign up to participate in a scavenger hunt. The sign-up sheet shows a total of 54 girls and 42 boys. The boys and girls will be divided into teams. Each team will have an equal number of boys, and each team will have an equal number of girls. There will be as many teams as possible. How many boys will be on each team? How many girls will be on each team? How many teams will there be in all?

NOTICE: Photocopying any part of this book is prohibited by law.

# 11. Home-School Connection

# Algebra

### Problem

Mr. Bemis hires a carpenter to build shelves and drawers in his storage closet. The carpenter charges $50 per hour for labor and $125 for materials. The carpenter charges Mr. Bemis a total of $300. How many hours does the carpenter work?

# 11. Home-School Connection

# Geometry

**Problem**

What is the sum of the angles in the figure shown below?

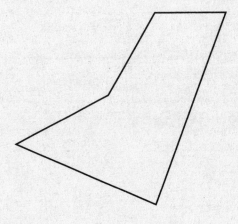

## 11. Home-School Connection

# Measurement

**Problem**

A hotel wants to apply 2 coats of paint to the walls and ceiling of a conference room. The room is 26 feet wide, 30 feet long, and 10 feet high. There are 4 windows, each measuring 6 feet by 3 feet, and two doors each measuring 7 feet by 3 feet that are not being painted. Each gallon of paint covers 400 square feet. How many gallons of paint should the painters order?

NOTICE: Photocopying any part of this book is prohibited by law.

# 11. Home-School Connection

# Data Analysis and Probability

**Problem**

A target consists of a 3-inch square that is centered within a 12-inch square. A dart is thrown at the target. Suppose the dart is equally likely to hit any point in the large square. What is the probability that the dart will hit the small square 2 times in a row?

# Glossary

# Glossary

## A

| | |
|---|---|
| **About** | In terms of quantity, a rough, estimated, or approximate number. |
| **Accurate** | Exact or precise. |
| **Additional** | In addition to; after the first. |
| **Algebra** | A type of mathematics that uses numbers, variables, and operations symbols to describe number relationships, rules, and patterns. |
| **Algebraic equation** | A mathematical sentence that includes one or more variables. Example: $n + 6 = 7$ |
| **Angle** | A figure formed by two rays that have a common endpoint, called the vertex. |
| **Apiece** | Each, or for each one. |
| **Area** | The number of square units needed to cover a region. |
| **Average** | See *mean*. |

# Glossary

## B

**Base of a parallelogram, triangle, or trapezoid**
A side of a parallelogram, triangle, or trapezoid, which includes one endpoint of the height.

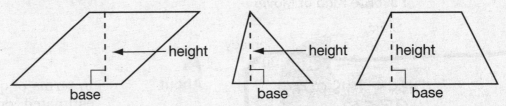

**Base of a solid figure**
A flat face of a solid figure; a prism is named by using the shape of its bases.

**Border**
The boundary of a region or a region that surrounds another region.

**Box-and-whisker plot**
A graph that uses a box to show the middle 50% of the data and two segments called "whiskers" to show the lowest 25% and the highest 25% of the data. Points on this graph include the lower extreme, the lower quartile, the median, the upper quartile, and the upper extreme.

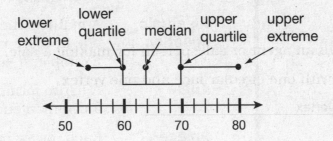

## C

**Center**
The middle point or part.

**Central angle**
An angle in which the vertex is at the center of a circle. A central angle creates a sector of a circle.

**Circle**
A closed two-dimensional figure having all points the same distance from a point called the center.

# Glossary

**Circle graph**  A graph in which a circle is used to represent a whole. Sections of the circle represent fractional parts of the data. Each section is a sector of the circle.

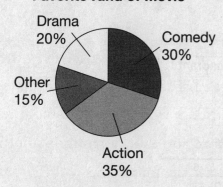

**Circular**  Having the shape of a circle.

**Circumference**  The distance around a circle.

**Combined**  Put together.

**Commission**  Amount paid to an agent or salesperson for making a sale.

**Cone**  A solid figure with one circular face and one vertex.

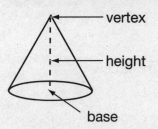

**Congruent**  Having the same size and shape.

NOTICE: Photocopying any part of this book is prohibited by law.

# Glossary

**Coordinate plane**  A system used to show location. It is formed by the intersection of two perpendicular number lines called axes. Points on the plane can be named by an ordered pair of coordinates, $(x, y)$.

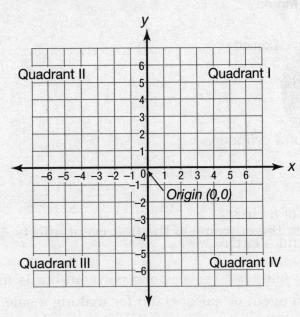

**Coordinates**  An ordered pair of numbers $(x, y)$ that identifies the location of a point on the coordinate plane

**Cost**  A certain price.

**Cube**  A solid figure with six congruent square faces.

**Cubic unit**  A unit used to measure volume, such as cubic centimeters or cubic inches.

# Glossary

**Cylinder**  A solid figure with two congruent circular faces and one face that is a curved surface that is also called the *lateral surface*.

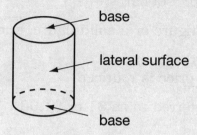

**Cylindrical**  Having the shape of a cylinder

# D

**Data**  Information.

**Dependent events**  Two events in which the outcome of the first event affects the outcome of the second event.

**Diagonal**  A line segment that joins two vertices of a polygon, but is not one of the sides of the polygon.

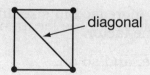

**Diameter**  A line segment that has passes through the center of a circle and that has endpoints on the circle.

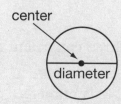

NOTICE: Photocopying any part of this book is prohibited by law.

# Glossary

| | |
|---|---|
| **Difference** | The answer in a subtraction problem. |
| **Dime** | A coin worth 10¢. It is worth $\frac{1}{10}$ of a dollar. |
| **Dimensions** | Linear characteristics of a plane figure or a solid figure, such as length, width, and height. |
| **Discount** | An amount by which an original price is reduced. |
| **Dollar ($)** | A bill or coin worth 100¢. It is referred to as $1 or $1.00. |

# E

| | |
|---|---|
| **Each** | Every one of a group considered individually. |
| **Edge** | A line segment that is the intersection of two faces of a solid figure, such as a prism. |

[figure: a rectangular prism with an arrow labeled "edge" pointing to one of its edges]

| | |
|---|---|
| **Enlargement** | An object such as a photograph that is increased in size. In an enlargement, each dimension is increased by the same factor. |
| **Equal** | Being the same in quantity, size, degree, and so on. |
| **Equally likely** | As probable as. |
| **Equation** | A mathematical sentence that says two numbers or expressions are equal. Examples: $2x = 8$, $2 + 2 = 4$. |
| **Equilateral triangle** | A triangle with three congruent sides and three congruent angles. |
| **Exactly** | Precisely. |

NOTICE: Photocopying any part of this book is prohibited by law.

# Glossary

**Excluding** — Leaving out.

## F

**Face** — A flat figure that is one side of a solid figure.

**Factor** — One of two numbers that are multiplied to give a product.
Example: In 5 × 6 = 30, 5 and 6 are factors.

**Fee** — An amount that is charged.

**Formula** — A statement that uses symbols to express a rule

**Function** — Two sets of numbers in which the input, $x$, is matched with exactly one number in the output, $y$.

## G

**Geometry** — The study of figures and location.

**Graph** — A picture that displays data.

**Greatest common factor (GCF)** — The largest number that is a factor of two or more given whole numbers.
Example: The GCF of 16 and 24 is 8.

**Guess and Test** — A problem-solving strategy in which you make a guess, test your guess, and then change your guess until finding the correct answer.

## H

**Height of a parallelogram, trapezoid, or triangle** — For a parallelogram or trapezoid, the perpendicular distance between the two bases. For a triangle, the perpendicular distance between a base and the opposite vertex.

**Height of a solid figure** — In a cone or a pyramid, the perpendicular distance between the base and the opposite vertex; in a prism or cylinder, the distance between bases.

**Hexagon** — A polygon with six sides.

NOTICE: Photocopying any part of this book is prohibited by law.

# Glossary

**Histogram**  A graph in which bars represent intervals of data.

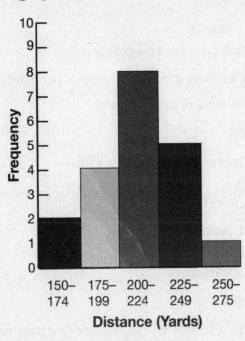

**Hourly**  A rate expressed per 60 minutes, or an event happening once every sixty minutes.

**Hypotenuse**  The side of a right triangle that is opposite the right angle.

# I

**Independent events**  Two events in which the outcome of the first event does not affect the outcome of the second event.

**Intersect**  To meet or cross.

**Isosceles triangle**  A triangle with two equal sides.

# Glossary

## L

**Lateral surface**  
In a cylinder, the curved surface.

**Least common multiple (LCM)**  
The least number that is a multiple of two or more given numbers. Example: The LCM of 6 and 8 is 24.

**Legs of a right triangle**  
The two sides that form a right angle in a right triangle.

**Length**  
The measurement of distance between two endpoints; a dimension of a two-dimensional figure or a solid.

**Less**  
Fewer.

**Long**  
Length from one end to the other.

**Lower extreme**  
The least value in a set of data.

**Lower quartile**  
The median of the lower half of a set of data.

## M

**Many**  
A large number.

**Mean**  
The sum of a set of data divided the number of values in the set.

**Measure of central tendency**  
A measure used to describe the "middle" of a set of data or a typical value in a set of data. The mean, median, and mode are measures of central tendency.

**Median**  
The middle number in an ordered set of numbers. If there is an even number of values in the set, the median is the mean of the two middle numbers.

**Minimum**  
Least possible number.

**Mode**  
The number that occurs the most times in a set of data. A set of data may have no mode, one mode, or more than one mode.

**Monthly**  
A rate expressed per month, or an event happening once a month.

**More**  
Greater in number or amount.

NOTICE: Photocopying any part of this book is prohibited by law.

# Glossary

**Much** — Large in number.

**Multiple** — The product of a whole number and any nonzero whole number.

Example: The multiples of 6 are 6, 12, 18, . . .

## N

**Nickel** — A coin worth 5¢.

**Number sentence** — An equation or inequality.

## O

**Octagon** — A polygon with eight sides

**Open-ended problem** — A math problem in which there is one correct answer, but more than one method of finding that answer.

**Ordered pair** — A pair of numbers (x, y) that gives the location of a point on a coordinate plane.

## P

**Parallel lines** — Lines in the same plane that never meet and remain the same distance apart.

**Parallelogram** — A quadrilateral in which opposite sides are parallel and congruent.

**Pattern** — A series of numbers or figures that follows a rule.

**Pentagon** — A polygon with five sides.

**Per** — For each.

Example: "Miles per hour" means miles for each hour.

NOTICE: Photocopying any part of this book is prohibited by law.

# Glossary

| | |
|---|---|
| **Percent** | A ratio of a number to 100; it is expressed using a percent sign (%). |
| **Perimeter** | The distance around a figure. The perimeter is found by adding the lengths of all of a polygon's sides. |
| **Perpendicular lines** | Two intersecting lines that form right angles. |
| **Point** | An exact location in space. |
| **Polygon** | A simple, closed figure formed by line segments. |
| **Prime factorization** | Expressing a number as the product of its prime factors.<br>Example: $24 = 2 \times 2 \times 2 \times 3$ |
| **Prime number** | A whole number greater than the number 1 that has exactly two whole-number factors: itself and 1. |
| **Prism** | A solid figure that has two congruent, parallel bases that are polygons. A prism is named using the shape of its bases. |

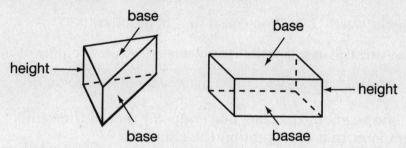

| | |
|---|---|
| **Probability** | The chance of an event occurring; it is equal to the number of favorable outcomes divided by the total number of outcomes. |
| **Product** | The answer in a multiplication problem. |
| **Proportion** | An equation that shows that two ratios are equivalent.<br>Example: $\frac{3}{8} = \frac{6}{16}$ |

NOTICE: Photocopying any part of this book is prohibited by law.

# Glossary

**Proportional** — Having equivalent ratios.

**Pyramid** — A solid that has one base that is a polygon; the rest of the faces are triangles that meet at a common vertex.

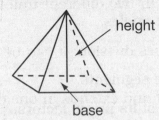

**Pythagorean theorem** — In any right triangle with legs $a$ and $b$, and hypotenuse $c$, $a^2 + b^2 = c^2$.

## Q

**Quadrant** — One of the four regions in a coordinate plane.

**Quadrilateral** — A polygon with four sides.

**Quarter** — A coin worth 25¢. It is equal to $\frac{1}{4}$ of a dollar.

**Quotient** — The answer in a division problem.

## R

**Radius** — A line segment that has one endpoint that is the center of a circle and another endpoint that is a point on the circle.

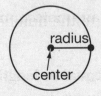

**Random** — Done by chance.

# Glossary

| | |
|---|---|
| **Range** | The difference between the greatest and least number in a set of data. |
| **Range of a function** | The possible values for the output, $y$, in a function. |
| **Rate** | A ratio that compares measurements in two different units, such as miles per hour or dollars per pound. |
| **Ratio** | A comparison of two numbers that uses division; it can be written as $\frac{a}{b}$, as $a:b$, or as $a$ to $b$. |
| **Ray** | A part of a line that has one endpoint and extends in one direction without end. |
| **Rectangle** | A parallelogram with four right angles. |
| **Rectangular** | Having the shape of a rectangle. |
| **Rectangular prism** | A prism in which all of the faces are rectangles (see *prism* for illustration). |
| **Reflection** | A transformation in which a figure is flipped over a line to create a mirror image. |

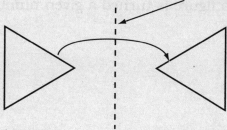

| | |
|---|---|
| **Regular price** | The price at which an item is regularly sold; an undiscounted price. |
| **Related** | Connected. |
| **Rhombus** | A parallelogram that has four sides that are equal in length. |

# Glossary

**Right angle**      An angle that measures exactly 90°.

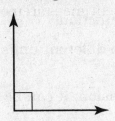

**Right triangle**      A triangle that has one right angle.

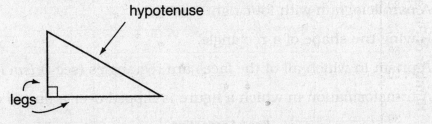

**Rotation**      A transformation in which a figure is turned a given number of degrees around a fixed point.

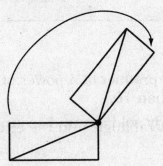

## S

**Same**      Equal or unchanged.

# Glossary

**Scatterplot**  A graph that consists of points (x, y) in the coordinate plane, in which x represents one measurement and y represents a different measurement.

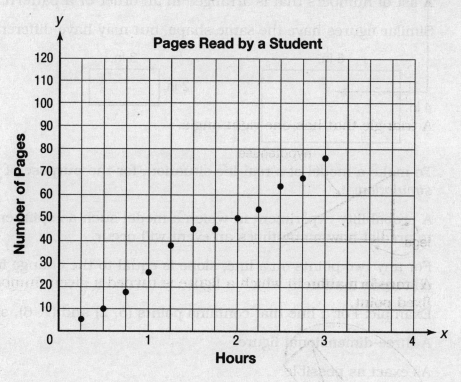

**Scientific notation**  A method of writing numbers as the product of a power of 10 and a decimal that is greater than or equal to 1 but less than 10.

**Sector**  A section of a circle created by a central angle and the circle.

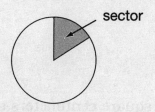

NOTICE: Photocopying any part of this book is prohibited by law.

# Glossary

| | |
|---|---|
| **Semicircle** | Half of a circle. |
| **Sequence** | A list of numbers that is arranged in an order or a pattern. |
| **Similar** | Similar figures have the same shape, but may have different sizes. |

6 in.

3 in.

2 in.

4 in.

| | |
|---|---|
| **Simulate** | To make a model of a real-life situation, for the purpose of making a prediction. See *simulation*. |
| **Simulation** | A probability experiment in which a model such as spinner or number cube is used to predict how many times an event will occur. |
| **Slope of a line** | For any two points on a line, slope is equal to the change in *y*-value divided by the change in *x*-value. |

Example: For a line that contains points (5, 2) and (7, 6), slope $= \frac{(6-2)}{(7-5)} = \frac{4}{2} = 2$.

| | |
|---|---|
| **Solid figure** | A three-dimensional figure. |
| **Specifically** | As exact as possible. |
| **Sphere** | A solid figure in which all points are equally distant from a point called the center. |

| | |
|---|---|
| **Square** | A rectangle with four equal sides. |
| **Square unit** | A unit used to measure area, such as square centimeters or inches. |

NOTICE: Photocopying any part of this book is prohibited by law.

# Glossary

**Stem-and-leaf plot**  A display in which data is organized into intervals; each piece of data is separated into a leaf that is the final digit, and a stem that represents the remaining digit or digits.

**Ned Kelly's RBI Totals**

| Stem | Leaf |
|---|---|
| 7 | 9 |
| 8 | 1 3 4 5 7 8 9 |
| 9 | 0 |
| 10 | 5 5 |
| 11 | 0 |

Key:  8 | 1  Means 81
      11 | 0  Means 110

**Sum**  The answer in an addition problem.

**Surface area**  The total area of the bases and other faces of a solid figure.

# T

**Total**  Sum or whole amount.

**Transformation**  A change that moves a figure. Reflections, rotations, and translations are transformations.

**Translation**  A transformation that slides each point in a figure the same distance in a given direction.

NOTICE: Photocopying any part of this book is prohibited by law.

# Glossary

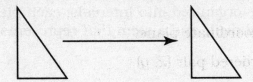

| | |
|---|---|
| **Triangle** | A polygon with three sides and three angles. |
| **Triangular** | Having the shape of a triangle. |
| **Triangular prism** | A prism in which the two bases are triangles. |
| **Triple** | Three times. |
| **Twice** | Two times. |

## U

| | |
|---|---|
| **Unit cost** | Cost per individual unit of an item. |
| **Upper extreme** | The greatest value in a set of data. |
| **Upper quartile** | The median of the upper half of a set of data. |

## V

| | |
|---|---|
| **Value** | Quantity or amount. |
| **Vertex** | A point where two rays meet; a point where two sides of a polygon meet. (plural, *vertices*) |
| **Volume** | The number of cubic units that a solid figure contains. |

## W

| | |
|---|---|
| **Width** | A dimension of a two-dimensional figure or a solid figure. |
| **Withdrew** | To take back or take away. |

# Glossary

## X

**x-axis**            The horizontal axis in a coordinate plane.

**x-coordinate**      The first number in an ordered pair $(x, y)$.

## Y

**y-axis**            The vertical axis in a coordinate plane.

**y-coordinate**      The second number in an ordered pair $(x, y)$.

# Math Abbreviations

| | | | | | | | |
|---|---|---|---|---|---|---|---|
| centimeter | cm | hour | h | milligram | mg | pound | lb |
| cup | c | inch | in. | milliliter | mL | quart | qt |
| day | d | kilogram | kg | millimeter | mm | second | s |
| fluid ounce | fl oz | kilometer | km | minute | min | ton | T |
| foot | ft | liter | L | month | mo | week | wk |
| gallon | gal | meter | m | ounce | oz | yard | yd |
| gram | g | mile | mi | pint | pt | year | y |

# Time Measurement

| Days of the Week | Months of the Year | Days in Month |
|---|---|---|
| Sunday | January | 31 |
| Monday | February | 28 or 29 |
| Tuesday | March | 31 |
| Wednesday | April | 30 |
| Thursday | May | 31 |
| Friday | June | 30 |
| Saturday | July | 31 |
|  | August | 31 |
|  | September | 30 |
|  | October | 31 |
|  | November | 30 |
|  | December | 31 |

# Larger Units of Time

1 week (wk): 7 days

1 month (mo): about 30 days

1 year (y): 12 months or 365 days

1 year: about 52 weeks

1 leap year: 366 days

1 decade: 10 years

1 century: 100 years

# Math Symbols

| | | | |
|---|---|---|---|
| + | addition | " | inches |
| − | subtraction | ' | feet |
| × | multiplication | °C | degrees Celsius |
| ÷ | division | °F | degrees Fahrenheit |
| = | is equal to | $\overleftrightarrow{AB}$ | line AB |
| > | is greater than | $\overrightarrow{AB}$ | ray AB |
| < | is less than | $\overline{AB}$ | line segment AB |
| . | decimal point | ∠A | angle A |
| $ | dollars | △ABC | triangle ABC |
| ¢ | cents | (2, 3) | ordered pair (2, 3) |

# Notes

# Notes

# Notes

# Notes

# Notes